EXPERIMENTALLY DETERMINATION OF THE INFLUENCE OF CORROSION ON THE MECHANICAL PROPERTIES OF STRUCTURAL STEEL S355JR

ANTONIO SHOPOV

2025

EXPERIMENTALLY DETERMINATION OF THE INFLUENCE OF CORROSION ON THE MECHANICAL PROPERTIES OF STRUCTURAL STEEL S355JR

Author: Dr. Antonio Shopov

Translated from Bulgarian

English, Second edition, Sofia, 2025

Editor: Dr. Svetlozar Ganev

Review: Assoc. Prof. Dr. Radi Hristov

CONTENTS

Chapter I. Literature review....9

1. Introduction....9

2. Corrosion of steels....9

2.1 Steel....9

2.2 Corrosion....10

2.3 Types of corrosion processes....11

2.4 Classification of corrosion effects on steels....12

3. Atmospheric corrosion....12

4. Accelerated corrosion....13

4.1 Salt spray corrosion....13

4.2 Cyclic corrosion test....14

4.3 Scrap test....14

4.4 Immersion in corrosive solution....14

4.5 Electrochemical method....14

5. Stress corrosion....15

6. Analysis of the regulatory codes....16

7. Effect of corrosion on material performance diagram....17

8. Effect of corrosion under cyclic loading....21

9. Effect of corrosion at high temperatures....23

10. Effect of torsional corrosion....23

11. Loss of resistance in the presence of corrosion....23

12. Modeling using the finite element method....24

13. Deterministic model....25

14. Stochastic model 26
15. Polyponal approximation 27
16. Summary 28
Chapter II. Materials and methods 29
1. Materials 29
2. Methodologies 30
Chapter III. Experimental part 32
III.A. Steel specimens 32
1. Basic setup 32
1.1. Dimensions of the steel specimen 32
1.2. Steel specimen material 33
1.3. Number of steel specimens 34
2. Electrochemical accelerated corrosion 36
2.1. Accelerated corrosion method for steel test specimens 36
2.2. Electrochemical corrosion of steel specimens 39
3. Cleaning the test bodies from corrosion products 41
4. Measurement after electrochemical accelerated corrosion 42
5. Processing of experimental data 43
III.B. Experimental study of mechanical properties 43
1. Study of steel specimens without corrosion 43
2. Study of steel specimens with corrosion 46
2.1. Formation of groups 48
2.2. Study of Group A 49
2.2.1. Elastic zone 49
2.2.2. Yield zone 50
2.2.3. Strain hardening zone 51

2.2.4. Necking zone 52

2.2.5. Stress-strain diagram 53

2.3. Study of Group B 54

2.3.1. Elastic zone 55

2.3.2. Yield zone 56

2.3.3. Strain hardening zone 57

2.3.4. Necking zone 58

2.3.5. Stress-strain diagram 59

2.4. Study of Group C 60

2.4.1. Elastic zone 61

2.4.2. Yield zone 61

2.4.3. Strain hardening zone 62

2.4.4. Necking zone 63

2.4.5. Stress-strain diagram 64

2.5. Study of Group D 65

2.5.1. Elastic zone 66

2.5.2. Yield zone 67

2.5.3. Strain hardening zone 68

2.5.4. Necking zone 69

2.5.5. Stress-strain diagram 70

2.6. Study of Group E 71

2.6.1. Elastic zone 72

2.6.2. Yield zone 73

2.6.3. Strain hardening zone 74

2.6.4. Necking zone 75

2.6.5. Stress-strain diagram 76

2.7. Study of Group F 76
2.7.1. Elastic zone 77
2.7.2. Yield zone 78
2.7.3. Strain hardening zone 79
2.7.4. Necking zone 80
2.7.5. Stress-strain diagram 81
2.8. Study of Group G 82
2.8.1. Elastic zone 83
2.8.2. Yield zone 84
2.8.3. Strain hardening zone 85
2.8.4. Necking zone 86
2.8.5. Stress-strain diagram 87
2.9. Study of Group H 87
2.9.1. Elastic zone 88
2.9.2. Yield zone 89
2.9.3. Strain hardening zone 90
2.9.4. Necking zone 91
2.9.5. Stress-strain diagram 92
2.10. Study of Group I 93
2.10.1. Elastic zone 93
2.10.2. Yield zone 94
2.10.3. Strain hardening zone 95
2.10.4. Necking zone 96
2.10.5 Stress-strain diagram 97
3. Analysis of results 98
3.1 Elastic zone 98

3.2 Yield zone ... 99

3.3 Strain hardening zone ... 99

3.4 Necking zone ... 99

3.5 Stress-strain diagram ... 99

4 Microstructural study ... 100

5 Summary ... 104

Chapter IV. Prediction of corrosion impact ... 105

IV.A. Theoretical modeling ... 105

1. Introduction ... 105

2. Model of the corrosion process ... 105

3. Forecasting the development of the corrosion process ... 106

3.1. Forecast ... 106

3.2. Types of models ... 106

3.3. Modeling ... 108

3.4. Forecasting the development of processes ... 114

3.5. Modeling a computing system ... 116

3.6. Creating a block-type algorithm ... 122

4. Forecasting in sustainable development ... 123

IV.B. Prediction of residual strength and relative deformation ... 125

1. Corrosion category C1 ... 125

1.1 Yield strength ... 125

1.2 Tensile strength ... 126

2. Corrosion categoryC2 ... 127

2.1 Yield strength ... 127

2.2 Tensile strength ... 128

3. Corrosion categoryC3 ... 130

3.1 Yield strength 130

3.2 Tensile strength 131

4. Corrosion categoryC4..132

4.1 Yield strength 132

4.2 Tensile strength 134

5. Corrosion category C5..135

5.1 Yield strength 135

5.2 Tensile strength 136

6. Corrosion categoryCX..137

6.1 Yield strength 138

6.2 Tensile strength 139

7. Analysis of results..140

Chapter V. Implementation of the research findings..141

Application 1. Application of the study to corroded reinforcing steel......................141

Application 2. Application of the study using a corroded steel truss for re-use.....141

Application 3. Application of the study in assessing the possibilities for continued of a corroded steel structure 142

References..143

Chapter I. Literature review

1. Introduction

There are many factors causing corrosion of metals, respectively there are many types of corrosion, which are generally divided into two groups - chemical gas corrosion and corrosion in non-electrolytes and electrochemical atmospheric, underground, microbiological, in crevices, sublayer, along the waterline, knife, layer, erosion, cavitation, fretting corrosion, corrosion fatigue, etc. all are listed in BDS ISO 8044-94. Corrosion changes the qualities of the surfaces and the structure of the surface layer, and also reduces the area of the load-bearing sections. As a result, it directly affects the strength-deformation properties. When a given product is subjected to mechanical stress, it is necessary to study the influence that corrosion exerts on the mechanical properties.

Depending on the aggressiveness of the environment, corrosion can occur at different speeds, and its influence on mechanical properties can manifest itself in specific ways as a function of time. When it develops under the influence of the usual working environment for the object, it is called operational corrosion. It is usually a process that occurs over time and its development can take tens of years. It is difficult to obtain a representative group of test specimens for mechanical tests. In order to facilitate the process of studying the influence on mechanical properties, it is preferable to accelerate corrosion by placing the materials in an environment that is significantly more aggressive than their natural working environment. Creating or maintaining the operational environment in laboratory conditions is difficult. Conditions that mimic operational ones are created, i.e. model corrosion.

Depending on the corrosion effect on the surface of the material, two types of corrosion, subgroups, are distinguished:

- general corrosion - covers the entire metal surface and can be uniform, uneven or selective (component-selective, structural-selective)
- local corrosion - affects only individual areas of the metal surface and can be spotty, ulcerative, pitting, intercrystalline, transcrystalline, or sublayer.

2. Corrosion of steels

2.1 Steel

Steel – is a metal, deformable alloy of iron with up to 2% carbon and other elements. By chemical composition, steels are divided into two groups – carbon and alloy [67].

a) Carbon steels – in addition to iron and carbon, they also contain so-called normal impurities such as: manganese (up to 1%) and silicon (up to 0.6%) and harmful impurities – sulfur (up to 0.07%) and phosphorus (up to 0.09%). Impurities are inevitably present in connection with its production. Depending on the carbon content, they are subdivided into:

- low carbon (up to 0.25%)
- medium carbon (from 0.25 to 0.60%)
- high carbon (from 0.60 to 2.00%)

b) Alloyed – in addition to the specified components, these steels also include alloying elements (chromium, nickel, molybdenum, tungsten, titanium, vanadium, and others), which are added in certain concentrations and increase the quality of the steel and give it special properties.

Depending on the purpose, steels are divided into the following main groups: structural, instrumental, special (with special physicochemical properties), acid-resistant, corrosion-resistant, fire-resistant, heat-resistant, electrical, etc. [67].

2.2 Corrosion

Corrosion is a process of destruction of steels due to the influence of aggressive factors. It is known that every substance tends to its initial state, since from an energetic point of view the form is the most stable. The corrosion process in iron is described by the equation [67] (Form. 1.1):

$$4Fe + 2H_2O + O_2 \rightarrow 4FeOH - q \quad (1.1)$$

During the process, energy q is released, which is used to produce iron [67]. Corrosion occurs slowly over time, and the released energy cannot be reused. The operation of buildings and facilities is constantly influenced by the environment - chemical, physical, biological, etc. During the process, steel begins to lose its properties until it is completely destroyed. The main factors that influence corrosion are: air, gases, water, acids, bases, temperature, solar radiation [67, 68, 71, 79].

Steels come into contact with certain substances from the air or water, undergo a chemical change that reduces the integrity of the steel. Oxygen, sulfur, salt and other materials increase the rate of corrosion. When the qualities of steel deteriorate, it cannot withstand the necessary impact as before corrosion. Structural steel used in bridges, railways and buildings is subject to corrosion. And it is necessary to monitor and manage corrosion to avoid an emergency. The corrosion rate is the rate at which any metal deteriorates in a certain environment and depends on the environmental conditions, as well as the chemical composition and condition of the steel. It is measured in millimeters that penetrate the cross-section each year. To calculate the rate, the following information must be taken into account:

- Weight loss (reduction in steel weight during the period)
- Density (density of steel)
- Area (total initial surface area of the steel part)
- Time (the duration of the action over a period of time)

The following formula (Form 1.2.) is usually used to calculate the corrosion rate:

$$[mm/y]=87,6*(W/(D.A.T)) \tag{1.2}$$

Where: W – mass loss in milligrams; D – density in g/cm^3 ; A – п cross-sectional area in cm^2 ; T time on corrosion influence in hours.

2.3 Types of corrosion processes

There is an enormous amount of research on the mechanism of the process, which has been summarized into two main types of corrosion - chemical and electrochemical.

a) chemical corrosion:

- gaseous, flowing in a gaseous medium
- corrosion in non-electrolytes, in liquids that do not possess the property of electrical conductivity

b) electrochemical corrosion – is characteristic of steels, the destruction occurs when an electric current occurs and flows from one area of the metal to other areas. This is due to the action of numerous microscopic galvanic elements formed on the surface of the metal.

- corrosion in electrolyte solution – occurs in natural waters and in aqueous solutions of electrolytes (acids, salts, bases)
- atmospheric corrosion – in atmospheric conditions and in any type of moisture-containing gas environment
- soil, ground or underground corrosion – when metals are exposed to soil or ground with a certain humidity, depending on the region
- electrocorrosion or corrosion from external current – caused by stray currents affecting metal
- contact corrosion – from contact of two different metals or alloys having different electrochemical
- stress corrosion – under the simultaneous influence of a corrosive environment and mechanical stress. There are: corrosion under periodically changing load, or corrosion fatigue, corrosion cracking from external tensile stress, corrosion cracking from internal stresses in the metal

- corrosion under simultaneous impact or abrasion of the external environment – it is called corrosion cavitation and is observed in pumps, propellers, hydroturbines, etc. Corrosion occurring under conditions of simultaneous abrasion of the environment is corrosion erosion.
- biocorrosion – cases of underground corrosion, or in electrolytes, when the acceleration of the process occurs with the participation of products released by microorganisms.

2.4 Classification of corrosion effects on steels

The corrosion category classification is based on the ISO 9223 standard, which deals with corrosion due to environmental influences. The classification is suitable for structural steels, as the environment affects them through corrosion development and is classified on the basis of loss per 1 year depending on the environment in which the structural element is located (Table 1.1).

Table 1.1 Corrosion categories depending on corrosion rate

Corrosion categories	C1	C2	C3	C4	C5	CX
Infulence	very weak	weak	average	highly	very highly	extreme
corrosion rate [mm/year]	< 0.0013	< 0.025	< 0.05	< 0.08	< 0.2	< 0.7

3. Atmospheric corrosion

The most common type of corrosion is the result of the combined action of moisture, temperature, solar radiation and environmental factors, forming a layer on the surface of metals, as a result of which they corrode [72, 78, 88]. An important factor for steel structures operating outdoors - bridges, buildings, power lines, pipelines, etc. The problem is significant for steel load-bearing structural elements, since steel corrodes easily as a result of environmental influences. Thus, the mechanical properties of steel structural elements decrease over time [72, 78, 88, 92, 93, 96].

Atmospheric corrosion is usually uneven and produces spots, ulcers and pitting – Fig. 1.1. Conventional methods for corrosion testing in sodium chloride solution give significantly better uniformity [71, 96]. This affects the mechanical behavior of test specimens. Corrosion testing methods must be chosen judiciously. There are corrosion methods that create accidents, but they are complex and require time, resources and effort [71, 96].

Figure 1.1 Influence of atmospheric corrosion on steel structures

Although atmospheric corrosion of steel structures is a slow process, its impact on mechanical properties is a current issue. Typically, test specimens are taken from structures that have been exposed to the environment causing the corresponding corrosion category for 20-40 years [1, 3, 4, 35].

To obtain the effects of atmospheric corrosion, an artificial atmosphere is also used. Despite the difficulties and the long duration of the study, the artificial atmosphere provides advantages in the preparation of groups of representative test bodies and is preferred as a basis for research. In the authors [3, 4, 35], despite the long duration of the corrosion effect (26 years in a chloride environment, under two types of load), there is a control group of bodies that have not been subjected to corrosion, with which a comparison is made. In experiments [20, 23, 39, 57], part of the steel structure is destroyed and samples with different degrees of corrosion are selected, which are tested and compared. In the processes, it is necessary to measure the corrosion time, the weight of the samples in order to accurately assess the degree of corrosion, after which a tensile test is performed and the results are compared. For test specimens with a complex shape, for example, ribbed reinforcing steel, the weight method is mainly applied and corrosion is given as a percentage of the original weight.

4. Accelerated corrosion

Accelerated corrosion is an artificial method of creating corrosion. There are five main methods of accelerated corrosion that are used [71].

4.1 Salt spray corrosion

The method with 5% NaCl solution (pH 6.5 ÷ 7.2) at a temperature of 35 ° C. The steel samples are placed in the salt spray chamber [15, 16, 95] in accordance with the ASTM B117-94 and ISO 9227 standards. The samples are removed from the chamber, cleaned with water and dried. The corrosion product, the oxidized rust layer, is then removed with a sharp brush. The presence of the oxide layer leads to a loss of weight from the sample. The weight loss is determined by the weight method. It has been found [10] that the loss increases with time. The stereoscopic test shows that after 10 days, the salt spray method, the surface will develop pitting corrosion. The weight loss is about 11% over a period of 60 days [7].

4.2 Cyclic corrosion test

This is a realistic accelerated corrosion method. A program-controlled chamber [48] is used to create combinations of cycles of parameters such as temperature, humidity, wet/dry conditions, solution concentration, and ultraviolet radiation. A study [9] of galvanized sheet metal subjected to 12 weeks of this type of method showed that the corrosion produced was equivalent to 1 year of tropical climate corrosion

4.3 Scrap test

The steel samples are placed outdoors at an angle of 45° facing south [29]. They are sprayed with a 3% NaCl solution [40, 43, 45] for a specified period of time. The test lasts for about 6 months. The test is standardized in ISO 11474:2014. It is an accelerated corrosion method that also includes wetting/drying cycles, changing solution concentration, temperature and ultraviolet solar radiation. The method is laborious and time-consuming, and therefore is not applicable to tensile strength tests.

4.4 Immersion in corrosive solution

It is convenient to implement from a technical point of view. The time to reach the desired degree of corrosion depends on the chemical composition of the solution and the corrosion resistance of the steel. In [11] natural seawater was used, the experiment lasted one year, the tensile strength was reduced by 14.5%. In [8, 12] 20% NaCl solution and flat test bodies with a cross section of 20 × 4 mm were used. In 8 days, a weight loss of 36 - 55% is achieved. Especially in cylindrical steel samples and in solutions close to the natural working environment, i.e. 3.5% - 5% NaCl [17, 30, 34, 92], the corrosion rate is too low. It is preferable to accelerate the process further, usually by passing an electric current through it.

4.5 Electrochemical method

The corrosion process is accelerated by using an external DC source and anodic dissolution of the steel [2, 24, 27, 41, 49, 81, 91]. The test specimens are connected to the positive pole of the source, and a stainless-steel plate or other inert metal is connected to the negative pole. A DC current of a certain value is applied for a certain time to achieve the degree of corrosion. The method is known as galvanostatic. A solution of 5% NaCl is applied [17, 19, 24, 49, 55, 81, 90, 91] and a current density of 0.1÷20 mA/cm2 [1, 17, 26, 31, 32, 42, 49, 50, 53, 54, 70, 81, 91]. In [5], a mass loss η of 2÷80% was achieved by applying a constant current I (A) for a time period t (s), determined by Faraday's law (form. 1.3):

$$t = \frac{\eta W z F}{M I}; \qquad I = A \times i \times 10^{-3} \tag{1.3}$$

Where: W e is the weight of the original steel specimen [g], *z* is the valence of the ion (z = 2.5 is the average value for Fe^{2+} and Fe^{3+} of the corrosion products), F = 96484 C/mol is the Faraday constant , M = 56 g/mol is the atomic mass of the ferric-ion , *A is the surface area,* cm^2. To ensure better accuracy, in the corrosion process the current I is checked to a value based on the current density and the residual surface area A (formula 1.4):

$$A = 2\sqrt{\frac{\pi l}{\rho} W(1-\eta)} \tag{1.4}$$

Where: ρ = 7,8 g/cm^3 is the density of the steel, l is the length of the test section, cm. The method allows the use of different NaCl solutions and different current strengths, as used by other researchers [63].

5. Stress corrosion

A form of corrosion that is a problem area in civil and industrial construction. It occurs when a material is in an inert environment but corrodes under applied stress. The stress may be applied externally or residually [87]. Corrosion has no visible factors prior to failure and can lead to catastrophic failure. It generally refers to the degradation and/or rust formation of a metal surface in an electrochemical environment subjected to tensile forces in a residual or direct form [87]. Stress corrosion erodes the microscopic granular composition of a metal surface, often causing cracking and disintegration of the surface [87]. Many steels alloys experience stress corrosion. Residual stress can be the stress that remains in the material after forming or stress due to welding. Stress corrosion cracking usually leads to material failure, which can have severe consequences, as the signs are almost non-existent before failure occurs [87]. This type of corrosion is in some cases a form of galvanic corrosion, where stressed areas of the material are anodized to non-stressed segments of the material [87]. In practice, the best way to control corrosion resistance is to limit or reduce the stresses at which the material is exposed to the corrosive atmosphere. The mechanism is by bending and the issue of stress relief is not addressed [87]. Areas of increased stress will be anodic to other areas. The outside of the element will have residual stress, as it is the part that will be elongated. The inside of the element will have residual compressive stress, as it is the side that will be contracted due to the bend present (Fig. 1.2).

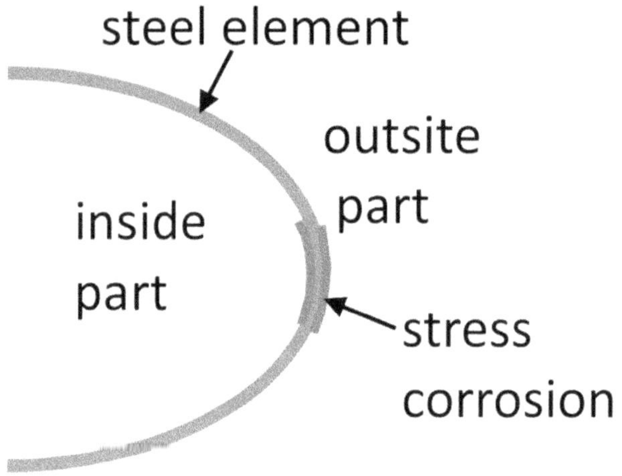

Figure 1.2 Scheme of the action of stress corrosion

6. Analysis of the regulatory codes

The following standards related to corrosion are currently in available:

- **BDS EN ISO 6892-1:2016 – Metallic materials. Tensile testing. Part 1: Test method at room temperature.** The test consists of loading a test specimen with a tensile force, usually until it breaks, in order to determine one or more of the following mechanical characteristics: tensile proportionality limit, tensile elastic limit, elastic modulus, lower and upper tensile yield limit, conditional tensile yield limit, tensile strength.
- **BDS 6372:1986 – Metals. Shear test method.** This test can be performed on all universal machines using a special device. During the test, the magnitude of the force is measured and recorded. The shear strength is determined.
- **BDS 1196:1971 – Metals. Torsion test method.** It is carried out with cylindrical solid or tubular test specimens on specialized machines. The following mechanical characteristics are determined: angular deformation modulus, relative torsional slip, torsional proportionality limit, conditional torsional yield strength, actual torsional strength limit; conditional torsional strength limit.
- **BDS EN ISO 7438:2016 – Metallic materials. Bending test.** A cylindrical or prismatic test specimen is loaded with a concentrated load according to two possible schemes. The maximum deflection in the middle of the test specimen and the magnitude of the force are measured.

- **BDS 5297:1983 – Metals. Fatigue testing methods.** The standard applies to methods for fatigue testing of test specimens made of metals and alloys, loaded in tension-compression, bending or torsion, under symmetric and asymmetric stress or strain cycles, in the presence or absence of stress concentration, at normal temperature and humidity according to BDS 15357-81. The purpose of the tests is to construct the August Wöhler diagram, also known as the S-N diagram or fatigue curve. Most of the initial fatigue tests are performed under a symmetric cycle R= –1 (asymmetry coefficient).
- **BDS EN ISO 6892-2:2011 – Metallic materials. Tensile testing. Part 2: Test method at elevated temperature.** The test consists of heating a test specimen to a specified temperature and loading it with a tensile force. The same mechanical characteristics are determined as at room temperature.
- **EN ISO 7539 "Corrosion of metals and alloys – Stress corrosion testing".** The tests may be carried out in tension or bending, on initially smooth test pieces or on those with a notch or pre-crack. A wide variety of test piece shapes and sizes may be used, the most commonly used being those described in EN ISO 7539-4 and EN ISO 7539-6. The test pieces are tested under constant load, increasing load or at low strain rates. The tests may involve bringing the test piece to complete failure and assessing the mode of failure in order to determine the susceptibility to stress corrosion cracking, or terminating the test at some intermediate stage and then determining the degree of crack initiation or growth. The assessment of the susceptibility to stress corrosion cracking is made by comparing identical test pieces exposed to corrosive and inert environments, applied for one or more of the following parameters: time to failure; ductility, assessed for example by reduction in area or elongation at failure; maximum load reached.

7. Effect of corrosion on material performance diagram

There is evidence that since 1967, studies have been carried out on the influence of corrosion on mechanical properties [22], two types of steels have been studied - St.2 and St.3, and it has been established that the influence consists of two main directions - reducing ductility and transforming the steel into a brittle material. In [1] it is stated that the degree of corrosion does not affect the tensile strength, if calculated with the actual cross-sectional area, as the study was done for reinforcing steel, and the degree of corrosion reaches 80%, the diameter starts from 11.75 mm and as a result of corrosion it drops to 4.50 mm. In [2], test specimens also made of reinforcing steel were tested, but with corrosion up to 17.19%. In the above studies, the disadvantage is that in three of the curves shown, the failure occurred outside the measuring length, especially in [2] an LVDT type extensometer with a working length of 50 mm was used, which did not record the entire diagram. The results show dispersion, since at 16 and 17% corrosion the strength is greater than at 6 and 15%. In [2] it is only commented that corrosion up to 15% has a relatively small influence on the deformation curve, while corrosion above 15% leads to a significant reduction in the plastic deformations and the residual capacity of the corroded bodies.

In [3], the effective (rather than nominal) cross-sectional area in the fracture was used to calculate the stresses. The measuring length was 200 mm, and the inductive extensometer was used. The conclusion is that in the corroded samples, the ductility decreases, but the type of steel is not noted, only that it is reinforcing steel. The conclusion acknowledges that the comparison of these parameters between samples with and without corrosion is complicated by the neck. The general conclusion is that the ductility in most of the corroded bodies decreases below the limit limited in the regulatory framework, and the residual deformation after failure in the corroded steel could be the most important parameter, as it affects the reliability of the steel structure. The dissertation [4] studies corroded reinforcing steel and the residual mechanical performance of corroded reinforced concrete beams. In a dissertation [14], investigating the effect of corrosion on the physical and mechanical properties of reinforced concrete, it was found that corrosion also affects it. Source [5] focuses mainly on the influence of the load during operation on the development of corrosion. It is shown that the presence of static load increases the depth and sharpness of the pits, and from there the strength decreases for reinforcing steel. In [6], cylindrical rods of A36 steel were used, from which test specimens of the tube type were obtained by turning in the measuring section, with a standard geometry according to the ASTM A-370 standard, and a universal tensile testing machine was used. Then, the test specimens were divided into three groups - control, 6 and 12 months of corrosion. The results show that the stresses decrease proportionally to the corrosion time.

In reference [7] presents mechanical tests of steel specimens of B420St steel subjected to accelerated corrosion, as well as specimens taken from real reinforced concrete beams from the Greek coast and from an industrial building, about 30-40 years old. [15] and [16] are related to [7], as B500St reinforcing steel was investigated. It was found that accelerated corrosion from 40 days of salt spray exposure is a realistic simulation of atmospheric corrosion of reinforced concrete from an old building on the sea coast. The effect of corrosion on the strength properties is most significant and leads to a reduction in the strength reserve (safety factor) during operation. The effect on tensile plasticity is significant. After 35 days of salt spray exposure, the elongation to failure falls below the 12% specified in the Greek standard. In [16], it was studied how it affects the deterioration of the mechanical properties of the reinforcing steel B500c in the presence of corrosion. It was found that at the same corrosion rate of 2.41%, the properties of the steel deteriorate more than the properties of the steel corroded directly in salt spray. Flat test specimens were used in [8], divided into groups, one of which was the control. The rest were immersed for different times (2-10 days) in a highly corrosive aqueous solution of 20% sodium chloride. A series of test specimens were made of three different steels - Iranian carbon St37, Ukrainian high-strength St52 and Japanese of the same brand St52. The sensitivity of the steels to corrosion based on their weight loss are identical up to 144 hours, after which it accelerates. The change in mechanical properties due to corrosion for all steels is the same and the tensile strength decreases, which must be taken into account when dimensioning structures for repeated use according to allowable stresses. In [12], the influence of pitting corrosion on the tensile deformation curve of the material was studied. Flat specimens were tested, with corrosion artificially produced using a machine. The same 3D models were used for numerical analyses using the finite element method with a program. Methods and formulas for determining the actual stresses and obtaining the actual working diagram of the material are given. An empirical formula is proposed for calculating the reduced strength and deformability depending on the magnitude of corrosion. It is shown that this reduction can be determined either by empirical formulas, or by simulation using the finite element method, or by using a strength correction factor, based on data from other studies [20, 27, 28].

In [17], the influence of corrosion on the mechanical properties of steel fibers was studied. Two corrosion methods were used – cyclic immersion/drying and galvanostatic method. In [27], a study was carried out, but with steel elements made of reinforcing steel, and an empirical formula was established. In [20], flat test specimens cut from a bridge exposed to atmospheric corrosion near the Pacific Ocean in Japan were studied after about 100 years of operation. The selected specimens contained three types of corrosion – general, pitting and local. They were tested in tension. Corrosion was studied with 3D scanning. It was found that the influence of corrosion on the working diagram of the material is significantly manifested around the yield point. A methodology was created for assessing the residual tensile strength by determining the minimum thickness of the cross-section, and a statistical method for prediction was also proposed. In [21], the influence of corrosion on the mechanical properties of B500D steel reinforcing bars under static and variable loading was experimentally investigated. It was found that the decrease in mechanical properties is not proportional to the degree of corrosion, but also depends on the influence of other factors, such as stress concentration at the tip of the pits, material inhomogeneity, displacement of the centers of gravity of the cross-sections from the longitudinal axis due to uneven loss of material during corrosion. In [24], reinforcing steel was subjected to accelerated electrochemical corrosion with a 5% sodium chloride solution. It was found that the strength of the reinforcement decreases with increasing corrosion. The elongation of the samples significantly decreases with corrosion and, moreover, the elongation of bars with corrosion of 15.6% or more is higher than that specified in IS: 1786-2008. Reinforcement with corrosion of 9.8% shows brittle behavior. A researcher has found [13] a reduction in ductility as a result of corrosion in Gr.70 steel according to ASTM A516 standard used for containers. There is a publication [69] in which the failure of cable wires of a bridge structure due to corrosion was investigated.

In [29], dependences for the degradation of the mechanical properties of corroded steel plates made of Q235 steel were developed using an experimental method and a numerical simulation method based on surface topography, and the results obtained partly overlap with [25]. In publication [35], reinforced concrete beams were exposed to salt spray (3.5% NaCl) for 27 years, after which the reinforcement, which was made of B420St steel, was removed and tested in tension. The loss of diameter was recorded, the strength was determined, and the loss of ductility was determined depending on the corrosion. In [54], reinforcing steel with corrosion in a reinforced concrete beam subjected to bending was investigated. The researchers in [39] established a numerical method, through an approximate first-order dependence of corroded steel plates, claiming that they established the dependence through an experiment, but there is no data on the type of steel, the way in which it corroded, as well as confirmation of the obtained result. [41], is a significantly detailed study in the field of the influence of corrosion on mechanical properties. A detailed map of India is given with marked zones of corrosion impact, 10 test specimens with different degrees of corrosion were studied in detail and it was found that corrosion negatively affects ductility, elasto-plastic properties, geometric characteristics, mechanical strength. As for profile tubular steel, it was found that with a percentage variation of corrosion from 20% to 40%, the maximum load capacity decreases from 10% to 25%, a significant drawback of this study is that it is not known what kind of steel it is made of. A group of researchers in [57], studied corroded test specimens made from the material of an existing bridge 75 years old. Their results found a loss of ductility and a change in the mechanical properties of the steel test specimens. The disadvantage is that they do not take into account that the existing anti-corrosion coating and this gives a reflection of the result. A study [62] of steel Q235 corresponding to S235, according to standard EN 10025, found a change in the yield zone due to corrosion, i.e. the yield point reaches a point where it disappears and brittle failure was found. There is a publication [66], which investigated the influence of local corrosion on short steel columns, which are used as bridge piles, found the degree of corrosion of the flange is a factor that has the most significant effect on the bearing capacity of the column. Other factors, such as asymmetry of the corrosion pattern or reduction of the flange width, have a negligible effect on the load-bearing capacity. Another publication [84] shows the influence that atmospheric corrosion has on existing steel river bridges in the Czech Republic, where corroded steel elements are attached to the bridge structure, the conclusion is that the first results are expected in 2026.

In [90], the mechanical behavior of corroded high-strength steel bars made of HRB500 steel under static and dynamic loading is investigated. Based on the experimental results, reduction factors are proposed to relate the tensile behavior to the degree of corrosion and the degree of deformation for the corroded bars. A modified Johnson-Cook strength model of corroded high-strength steel bars under dynamic loading is proposed, taking into account the influence of the degree of corrosion. Study [94], investigated different types of reinforcing steel B500c, B450c, B400c and B500b. The authors concluded that structural problems arising from the corrosion process are a common phenomenon for steel bars, which is mainly related to the formation of microcracks, pitting is due to the various aggressive factors that exist in any corrosive environment and lead to the deterioration of existing damage both superficially and deeply. Research [95], investigates the corrosion behavior and mechanical properties of steel Q345 corresponding to S355 according to standard EN 10025, by tensile testing, presenting the process of initial corrosion and analyzing the characteristics of the lower surface and the upper surface of the corroded sample. The stochastic distribution function and the dependence on time and the mass loss coefficient are presented. The influence of corrosion on the ultimate elongation, ratios, elastic modulus, stresses and ultimate strength of the studied steel is established. A significant drawback, as the authors themselves conclude, is that the corrosion process, dependent on time, could be many times better than the model they used, taking into account how the degree of corrosion is reflected by the mass loss coefficient, and from there on the mechanical properties. They reached a maximum mass loss of a test body of 2.08%, did not establish the presence of brittle fracture, did not study the change in mechanical properties over time and/or depending on the ongoing corrosion effect, i.e. long-term.

8. Effect of corrosion under cyclic loading

In [21], the influence of corrosion on the mechanical properties of B500D steel under static and variable loading was experimentally investigated. The influence of pitting depth on the number of cycles to failure varied from 8000 to 10000, while the diameter of the elements was 10 mm and 12 mm, the dependence between the degree of corrosion and Np was very significant and could be given by an exponential law. In [28], the authors proposed a formula for determining the change in E-modulus as a function of the changed diameter due to corrosion, which formula is linear, and the other proposed formula for calculating the stress due to corrosion is a quadratic function for AH32 steel used mainly in shipbuilding. In [33], the residual fatigue strength in corroded steel beams of a bridge structure composed of welded 2T beams made of A588 steel is considered. The method takes into account the reduction of the cross-section, the humidity of the environment and the concentration of stresses in the corroded pits. These parameters can be studied separately or in any combination. Numerous fatigue curves are given [28, 33, 46, 50]. Machine elements suffer from degradation of steel caused by corrosion. From the studies, relationships between "fatigue - stress" can be established [40, 42, 43, 48]. Relationships between "number of cycles" and "crack formation" are given [44]. Estimates of energy density dissipation have been made, depending on the number of cycles and the deformations that have occurred and the change in the rectilinear axis of the element [46, 50]. In order to perform fatigue tests, dynamic loading should also be applied. In [40] the dynamic loading is with a frequency of 95 Hz (5700 rpm) for $5x10^6$ cycles applied to AISI 4140 steel. In other sources a different dynamic loading is used - 10 Hz for 10^8 cycles and stainless-steel SUS 316L [42]. Frequency 50 Hz, until the element breaks at the corresponding tensile force (270, 333, 382, 449 and 515 MPa) of AISI 4340 steel [43]. Tensile force of 1610 MPa and dynamic loading with a frequency of 5 Hz, BS250 A53 steel [44]. Dynamic loading with a frequency of 30 Hz, on aluminum alloy 7075-T6 [48] and by using a machine [45, 49]. Orthopedic implants are made of 316L steel, which is stainless, but a study in [47] found that pitting corrosion occurs as a result of the saline solution that is infused into a person's body and, taking into account the cyclic loading of the implant, there is a loss of load-bearing capacity. In [50] using a force of 250 kN with a constant strain of 0.0003 strain/second, with a diameter of 12 mm from B500 steel. Dynamic loading with a machine with a frequency of 1 Hz for 20 hours, for 28 days [55]. By twisting 5x105 cycles, AISI 1020 low carbon steel was used, and a microstructural analysis was also made [56]. Dynamic loading with a frequency of 50 Hz, microstructural analysis, which was used AISI 409 SS steel [63]. In [82] a study was made with a dynamic loading with a machine with a frequency of 0.5 Hz, S400 steel, corresponding to S275 according to standard EN 10025. In publication [92] a study of S235JR steel is presented. The results obtained show a slight improvement in the mechanical and operational properties of the steel in corrosion conditions under constant load. Under variable loads and service corrosion, the results of steel fatigue do not confirm the beneficial effect.

The review in this area found that the studies carried out to date represent empirical dependencies and are linked to the development of corrosion, such as changes in geometric characteristics and in particular the structural changes it causes, but are not tied to time.

9. Effect of corrosion at high temperatures

In practice, many elements are operated at temperature ranges from -30^0 C to +50^0 C, but there are also extreme situations, such as fire. In these situations, the corroded element actually performs its functions at high temperature. Studies have been carried out [31, 32, 34], where the mechanical characteristics of corroded elements subjected to temperatures from 120^0 C to 1000^0 C were determined. High temperature tests are carried out in a different range. In [31] the temperature range is from 850^0 C to 1000^0 C, and cylindrical test specimens with a diameter of 4 mm made of AISI 318 LN steel according to ASTM E9 standard were tested. Temperature range from 400^0 C to 1000^0 C, the experimental specimen is a steel plate with dimensions 10x10x55 mm from UNS S31803 steel, using an INSTRON Wilson-Wolpert Tukon 2100B machine [32]. Practical temperature range is from 120^0 C to 195^0 C, with an INSTRON 4507 apparatus, for aluminum alloy AA7150-T6 [34]. The sterilizer operates at a temperature of 120^0 C to 250^0 C, the solution starts corrosion formation processes [53]. Studies have been conducted at temperatures from 150^0 C to 1500^0 C [70].

The studies have established that hardness and strength increase, but elasticity and impact toughness decrease. A change in the microstructure of the steel is established. No formulas have been derived from the presented results, no dependencies have been sought, both in terms of the development of corrosion and the influence it has on stresses and relative deformation.

10.Effect of torsional corrosion

In a study [75], a torsional test was performed on a machine element with pitting corrosion present. The results found that corrosion was identified as the main root cause of premature shaft failure. Corrosion-induced pitting would lead to very high glucocorticoid stresses and facilitate fatigue crack initiation, while the effect of corrosion would significantly reduce the durability and limit of the shafts under torsional loading.

11.Loss of resistance in the presence of corrosion

According to [2], the behavior of corroded reinforcing steel under compression and cyclic loading is important for the behavior of the entire reinforced concrete structure under earthquake conditions. They studied the influence of corrosion on buckling at 9-10 cycles. It is mentioned that until 2012 there were no studies on buckling of corroded steel in the plastic range, and this type of tests were also performed, and results were given, through which experimental dependencies were established between the weight loss due to corrosion and the stress together with the deformation. In [51], compressed pipes were studied and the relationship between the loss of wall thickness of a steel pipe caused by corrosion and the compressive force was analyzed. Change of axis and the appearance of eccentricity due to corrosion, and the change in the ratio of diameter to thickness of the section of the steel pipe was studied. In study [61], a study was made on the influence of corrosion on the stem of a steel beam and its loss of strength. In studies [73] on the loss of strength of a stem and/or a compressed element. In publication [83], a study was made on corroded angle steel profiles. Corrosion can be described by four main parameters – reduction in section thickness, corrosion spread, corrosion location and length of the corroded part. In [10], an experimental and numerical analysis of corroded steel plates subjected to compression and loss of strength was made. A formula for calculating the reduced buckling force was proposed, since it was found that the Euler formula gives inflated values. In [11], an analysis was made with a finite element program of corroded bars with L-shaped cross-section subjected to compression and loss of strength. The types of buckling and failure at different b/t ratios of the section and different corrosion positions are studied. In [18], 24 pcs. bars with L-shaped cross-section, corroded to different degrees by the galvanic method, are tested for compression and buckling. The load capacity of the bars is calculated in three theoretical ways, after which it is compared with the experimental results. The comparison shows insufficient accuracy of the calculation procedures.

In [39], issues such as [20, 23], concerning atmospheric corrosion, bridge construction, flat test bodies, 3D-scanning, FEM modeling are treated. The load is compressive and the residual critical force at loss of stability is studied.

12.Modeling using the finite element method

In source [5], corrosion modeling was done using the finite element method. Three trivial conclusions are made: that with increasing depth and width of the pits, stresses and deformations increase; that depth has a greater influence than width, that pits with sharp edges and peaks are the most dangerous. In publication [9], using the finite element method, with the help of a program, material deformation curves were obtained in pure shear of ideal plates in the presence of pitting corrosion. Plates of different thicknesses, with different arrangement, density and thickness of the pits were modeled. Publication [23] is related to [20] and presents an extremely successful modeling of the effects of corrosion on mechanical properties. A simple and accurate analytical model for estimating residual strength by measuring the maximum depth of corrosion is proposed. In [36], the influence of pitting corrosion on compressive strength was studied. A plate is modeled in a program, with indentations of different length, width, depth and position. The conclusions are trivial. An empirical formula is proposed for predicting the reduction in compressive strength, according to corrosion parameters.

In publication [73], a finite element model was developed after conducting the relevant tests. In publication [74], the researchers modeled the development of pitting corrosion under torsion using the finite element method. In publication [77], a realistic model for simulating the effects of corrosion on a steel pipe is presented. The interesting thing about this model is that first the pipe is taken, then a mesh is placed in the places where corrosion will occur, and then the model is built.

13.Deterministic model

In mathematics, computer science, and physics, a deterministic model is a system in which no randomness is involved in the development of the future states of the system [58, 98]. Therefore, a deterministic model will always produce the same result from a given initial state or initial condition [98]. In physics, laws that are described by differential equations are deterministic systems, although the state of the system at a given time may be difficult to describe explicitly. In mathematics, systems that are studied by chaos theory are deterministic [98]. If the initial state is precisely known, then the future state of such a system can theoretically be predicted [98]. In practice, however, knowledge of the future state is limited by the precision with which the initial state can be measured, and chaotic systems are characterized by a strong dependence on the initial conditions. In computer science and technology, a deterministic model of computation is one in which the successive states of a machine are actually the same, and the operations to be performed are completely determined by the previous state [98]. A deterministic algorithm is an algorithm that, given a particular input, will always produce the same output, with the underlying machine always going through the same sequence of states [98]. It is possible to implement non-deterministic algorithms that run on a deterministic machine, such as: an algorithm that relies on randomly selected numbers. Typically, a pseudorandom number generator is used for such random selections, but some external physical process is also used, such as the last digits of the time given by a computer clock [98]. A pseudorandom number generator is a deterministic algorithm that is designed to produce sequences of numbers that behave like random sequences [98].

This type of model cannot be applied as reliable because corrosion indicates an impact dependent on the random development of unrelated quantities.

14.Stochastic model

In probability theory, a stochastic process is the opposite of a deterministic process. In this method, the related fields stochastic (random) is a process in a mathematical object, which is usually defined as a collection of random variables [52, 58, 59, 60]. Historically, random variables have been related or indexed by a set of numbers, which are usually viewed as points in time, giving the interpretation of a stochastic process as representing numerical values of some system that changes with time [52, 58, 59, 60]. Stochastic processes are widely used as mathematical models of systems and phenomena that appear to be random. They have applications in many disciplines, including sciences such as biology, chemistry, ecology, neuroscience, and physics, as well as technological and engineering fields such as image processing, signal processing, information theory, computer science, cryptography, and others [52, 58, 59, 60]. The term random function is also used to refer to a stochastic or random process, because a stochastic process can also be interpreted as a random element in a function space. The terms stochastic process and random process are used interchangeably, often without a specific mathematical space for the set that indexes the random variables [52, 59, 60]. However, these two terms are often used when the random variables are indexed by integers or an interval of the real line. If the random variables are indexed by the Cartesian plane or some higher-dimensional Euclidean space, then the collection of random variables is usually called a random field [52, 59, 60]. The values of a stochastic process are not always numbers and can be vectors or other mathematical objects. Based on their mathematical properties, stochastic processes can be divided into different categories, which include random walks, martingales, Markov processes, Lévy processes, Gaussian processes, random fields, renewal processes, and branching processes [52 , 59 , 60].

The study of stochastic processes uses mathematical knowledge and techniques from probability, calculus, linear algebra, set theory, and topology, as well as branches of mathematical analysis such as real analysis, measure theory, Fourier analysis, and functional analysis [52, 59, 60]. The theory of stochastic processes is considered an important contribution to mathematics and continues to be an active topic of research for both theoretical and applied reasons. Instead of dealing with only one possible realization of the process in time (as in the case of, for example, solutions to an ordinary differential equation), in a stochastic (random) process there is an uncertainty about its future development (evolution), described by probability distributions [52, 58, 59, 60]. This means that even if the initial condition (or starting point) is known, there are many possibilities for how the process may unfold, with some realizations being more likely than others [52, 59, 60].

For corroded steel, the study of certain parameters implies that they are independent random variables [89], which dictates that the data processing model should be the stochastic method.

15.Polyponal approximation

In mathematics, approximation theory is concerned with how best to approximate functions by simpler functions and with quantifying the errors introduced by them [64, 80, 85, 86]. Recognizing that what is meant by best and simpler will depend on the application. A closely related topic is the approximation of functions by generalized Fourier series, i.e. approximations based on the summation of a series of terms based on orthogonal polynomials [64, 80, 85, 86]. One problem of particular interest is the approximation of a function using operations with the aim of approximating as close as possible to the actual function, usually with an accuracy close to that of floating-point arithmetic on the fundamental parameter [64, 80, 85, 86]. This is achieved by using a high degree polynomial and/or narrowing the domain over which the polynomial must approximate the function [64, 80, 85, 86]. Polynomials are functions with useful properties. Their relatively simple form makes them ideal for use as approximations for more complex functions [64, 80, 85, 86].

The polynomial in f(x) is a function of the form [64, 80, 85, 86] (Form. 1.5):

$$f(x) = a_0 + a_1.x + a_2.x^2 + a_3.x^3 + a_4.x^4 + ... + a_n.x^n \quad (1.5)$$

If we consider the case where, instead of knowing an expression for the function, we have values at points [64, 80, 85, 86]. It is enough to find a polynomial that passes through these points, and we want the polynomial to pass exactly through the given data, i.e. an interpolating polynomial [64, 80, 85, 86].

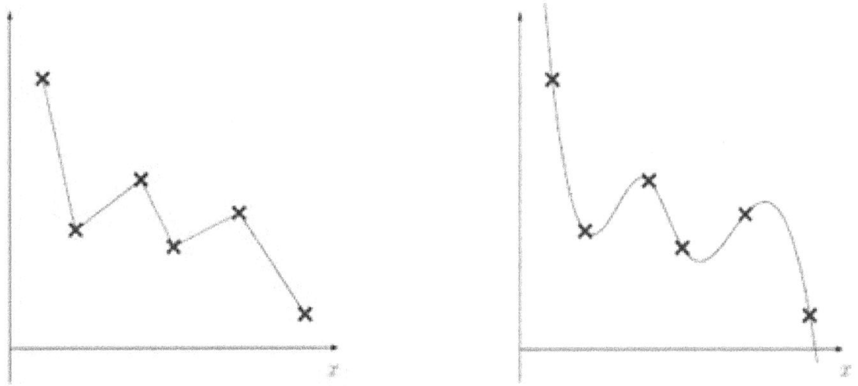

(a) (b)

Figure 3.21 (a) Linear connection on the "point-to-point" principle [64, 80]; (b) "Point-to-point" connection via function (no "corners") [64, 80]

Let us assume that we know (or choose to try) a function f(x) at exactly a few points and that we want to approximate the behavior of the function between these points [64, 80, 85, 86]. In its simplest form, this is equivalent to linear fitting (Fig. 3.21(a)), but it is often more accurate to look for a curve that has no "corners" in it (Fig. 3.21(b)) [64, 80, 85, 86].

In case we have experimental data after conducting an experiment and after making a line (function) passing through as close and accurate as possible to the data that was obtained [64, 80, 85, 86].

16.Summary

In the literature review, there is no information on research related to the prediction of the influence of corrosion on the mechanical properties of S355JR steel. The cited sources show different approaches and diversity in the search for specific solutions on the influence that corrosion has on the mechanical properties, but there are no scientific reports on the application of computational models for predicting the corrosion influence on steel structural elements.

A large number of studies have been carried out on corrosion. In practice, the characteristics of S355JR steel have not been comprehensively studied with a view to its utilization and possibilities for its practical application.

There are literary data on the analysis of corroded steel, but no analyses were found taking into account the influence of corrosion.

No system for accelerated corrosion has been developed that would include theoretical and practical development of S355JR steel

There are no scientific reports on the influence of corrosion over time and what impact it has on the mechanical properties of steel elements. No studies have been conducted on the influence of corrosion for S355JR steel, which is a structural steel.

There are no reports of comparisons of corrosion with real mechanical tests, there is no database on corrosion, numerous methods and stimulation models have been developed to determine the influence of corrosion on structural elements, but they require unjustified costs of time and funding for their implementation, and a low level of reliability of the results, both for steels and specifically for S355JR steel.

Chapter II. Materials and methods

1. Materials

- Steel S355JR, according to standard EN 10025:1:2004 "Hot-rolled structural steel products" from "ThyssenKrupp - Jupiter Steel" Ltd. (Bulgaria)
- 5% sodium chloride solution – prepared by adding 58.4 g/L sodium chloride (NaCl) to distilled water, stirring until sodium chloride is completely dissolved, the finished solution is manufactured in accordance with ISO 3696 / BS3978 standards, using high purity salts, deionized water, certified solvents, analytical balances and Class A glassware in a controlled environment, controlled with certified thermometers. Manufactured by Hanna Instruments (USA).

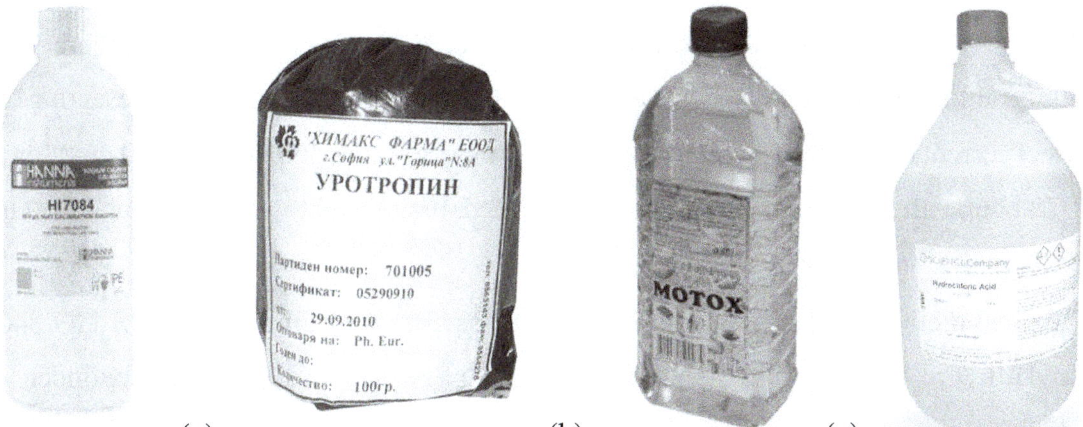

(a) (b) (c) (d)

Figure 2.1 Materials used – (a) 5% sodium chloride solution; (b) Dry alcohol; (c) Distilled water; (d) Hydrochloric acid.

- Distilled water - water that has been purified from all impurities through the distillation process. Manufactured by Aromatic Ltd. (Bulgaria)
- Hydrochloric acid – usually produced by the reaction of hydrogen with chlorine, after which the hydrogen chloride is dissolved in water, and the other way is the reaction of table salt with sulfur dioxide, standard ISO 905:1976. Manufactured by The Scince company (USA)
- Dry alcohol (hexamethylenetetramine) - is a chemical substance obtained from the reaction between 6 moles of formaldehyde and 4 moles of ammonia, produced by the company "Himax Pharma" EOOD (Bulgaria)
- Electric current – 220V , 50 Hz

- Current stabilizers model LM317 – manufactured by STMicroelectronics (Switzerland)
- Plastic box – 145 liters, made of polymer materials. Manufactured by Regalux GmbH (Germany)

(a) (b)

Figure 2.2 Materials used – (a) Current stabilizer LM317; (b) Plastic box - 145 liters

2. Methodologies

- Electrochemical accelerated corrosion – the test specimens are placed in a box (bath) with a 5% sodium chloride solution, connected to the current stabilizer separately and representing an anode. Direct current flows through them (100 mA, 200 mA, 300 mA, 400 mA and 600 mA) for a period of 72 hours, after which they are removed, cleaned, their weight is measured, and the change in cross-section (diameter) within 24 hours.
- Chemical method for cleaning corrosion products (rust) according to the BDS ISO 8407:2014 standard. The procedure is: the corroded test body is immersed in a glass cylinder with the solution, suspended on a copper wire and left in the solution for 10 minutes. After which it is removed and, if necessary, cleaned with a brush.
- Stochastic method - In the early 1930s, Alexander Khinchin gave the first mathematical definition of a stochastic process as a family of random variables indexed by the real line [52, 59, 60]. In experiments, it is a mathematical object that is defined as a collection of random variables.
- Polynomial approximation - In mathematics, approximation theory is concerned with how functions can best be approximated by simpler functions and with quantifying the errors introduced by them [80]. A closely related topic is the approximation of functions by generalized Fourier series, i.e., approximations based on the summation of a series of terms based on orthogonal polynomials.

- Mechanical tensile tests according to standard BDS EN ISO 6892-1:2016 with a universal tensile testing machine model BETA200-7/6x14 (Fig.2.3(b)) from Messphysik Materials Testing GmbH (Germany).
- Scanning electron microscopy using SEM model JSM 5510 (Fig. 2.3(a)) from JEOL Ltd. (Japan).
- Determination of the chemical composition using a spectral device "Spectrotest – TXC02; S/N 4S0007; analyzer: Fe-01MO; standard sample: RH 31/12, S3-2; type of probe: SL75060798 from SPECTRO Analytical Instruments GmbH (Germany).
- Determination of elemental composition using an optical emission spectrometer SPAS-02 from Active Co. Ltd. (Russia).

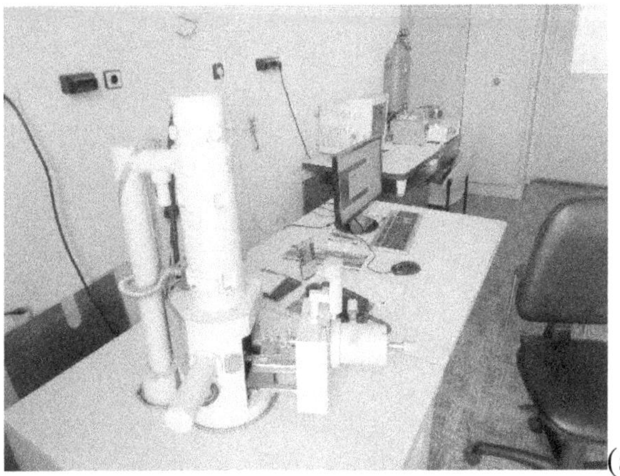
(a)

(b)

Figure 2.3 (a) Scanning electron microscopy model JSM 5510; (b) Tensile testing machine model BETA200-7/6x14

Chapter III. Experimental part

III.A. Steel specimens

1. Basic setup

1.1. Dimensions of the steel specimen

The selection of a steel test specimen is a complex task and depends on the standard to which it will be tested, the type of test, the cross-section of the test specimen, etc. Researchers use a variety of test specimens of various types and shapes [8, 12, 29, 57, 62, 65, 90, 95]. Shown in Fig. 3.1 – different types:

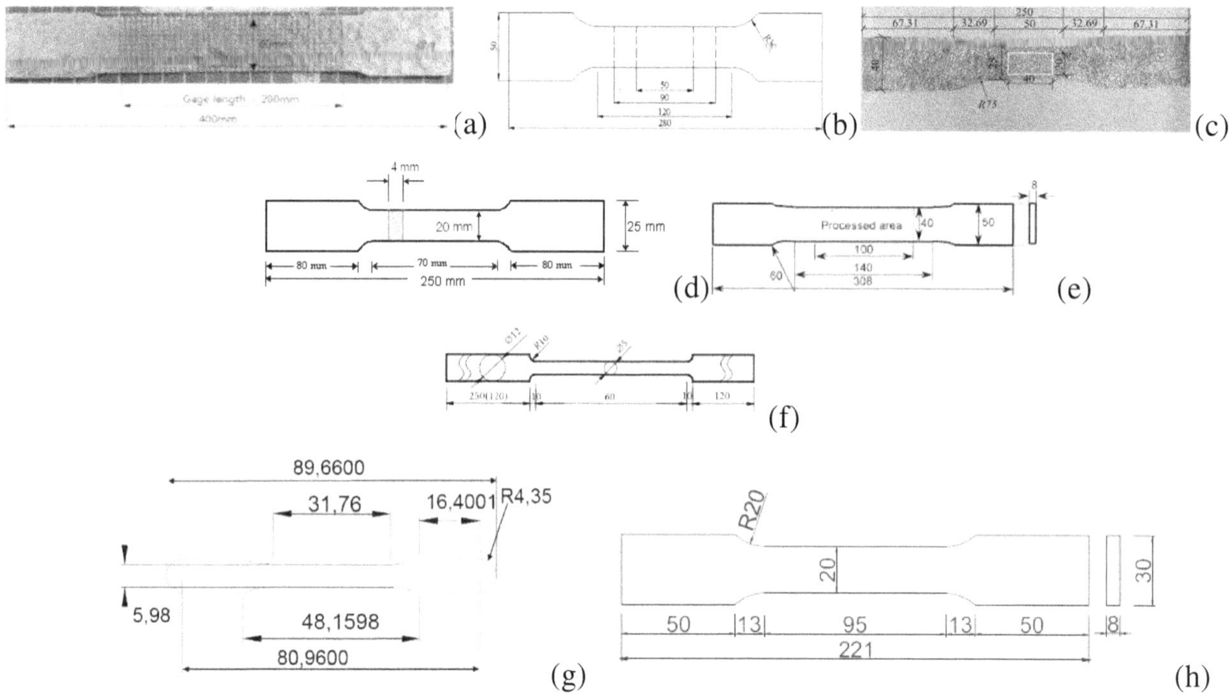

Figure 3.1 Dimensions and type of steel specimens
– (a) [57]; (b) [62]; (c) [29]; (d) [8]; (e) [12]; (f) [90]; (g) [65]; (h) [95]

Using the theory of strength of materials, it is recommended that the parallel length of the test piece be 15d [75, 76], and in [76] it is argued that this length avoids the influence of stress concentration in the areas of variable cross-section. The standard for tensile testing is ISO 6892-1:2016, which specifies the minimum parallel length of the test piece, but not the maximum length (Fig. 3.2):

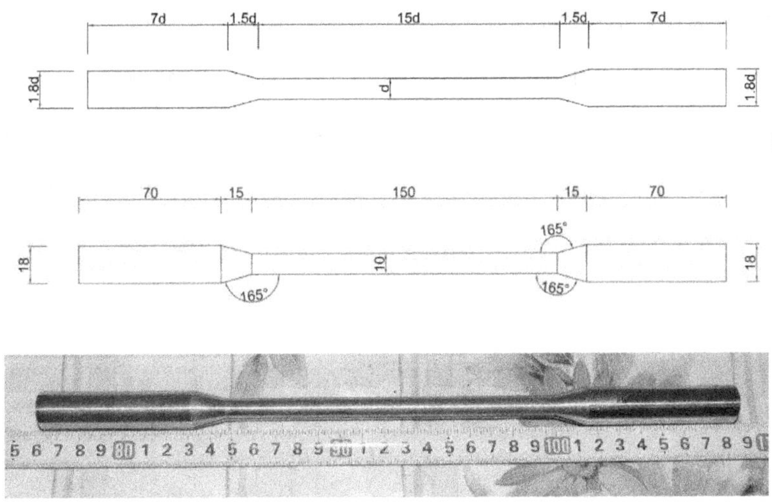

Figure 3.2 (a) Principal ratio of the dimensions of a test body as a function of diameter in the parallel length [49]; (b) Dimensions of the steel specimen [81, 91]; (c) Photo of a steel specimen according to dimensions made of S355JR steel [49]

1.2. Steel specimen material

According to the standard EN 10025:1:2004 "Hot-rolled products of structural steels", the types of steel that can be used as structural are defined. The mechanical properties of structural steels are the subject of various studies, but it has been established that steel S355JR has not been studied. Due to the presence of the chemical element silicon-Si up to 0.55%, which directly affects the strength of the steel. By increasing the strength, but reducing the elongation and transverse contraction, and resistance to corrosion. This steel should be tested for changes in its mechanical properties due to corrosion. According to the standard, the chemical composition of steel S355JR is (table 3.1) and the yield strength (minimum) $fp_{0.2}$ = 355 MPa.

Table 3.1 Chemical composition of steel S355JR, according to standard EN 10025:1:2004

chemical element	C carbon	Mn manganese	Si silicon	P phosphorus	S sulfur	N nitrogen	Cu copper	CEV others
max. percentage content, [%]	0.240	1.600	0.550	0.040	0.040	0.012	0.550	0.450

1.3. Number of steel specimens

125 steel specimens were made of S355JR steel. Of these, 116 test specimens were corroded using the accelerated corrosion method, 8 test specimens were tensile tested without corrosion and 1 test specimen without corrosion was kept as a control. Before being subjected to accelerated corrosion, the weight of each specimen was measured using a KERN scale, model PMJ 600 – 3M, with an accuracy of 0.001 grams (Fig. 3.3)

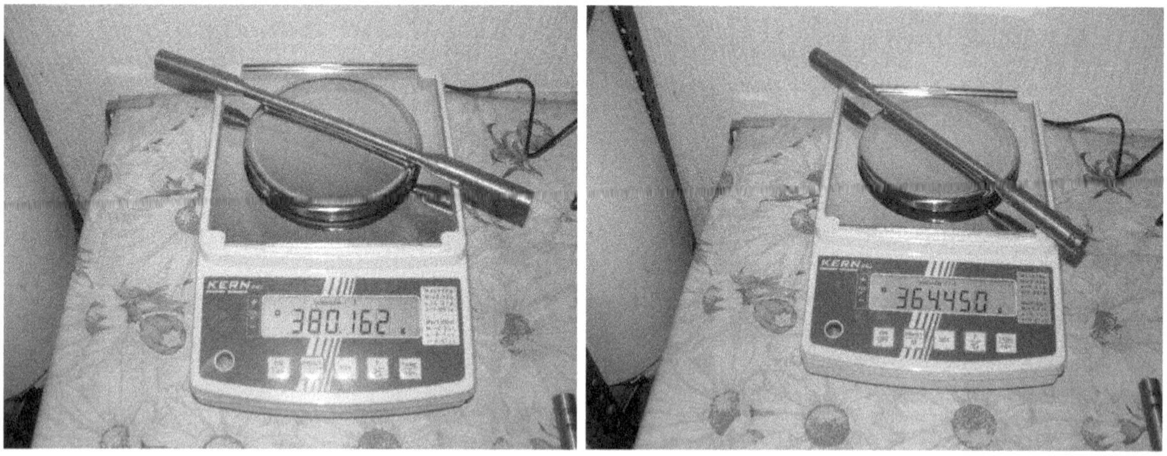

Figure 3.3 Moment of initial measurement of steel specimens

The steel specimen (test body) is stamped with a number at both ends, so that it is uniquely identified. Divided into 5 corrosion groups, at current values - 100 mA, 200 mA, 300 mA, 400 mA and 600 mA. The number of each group is as follows (Table 3.2):

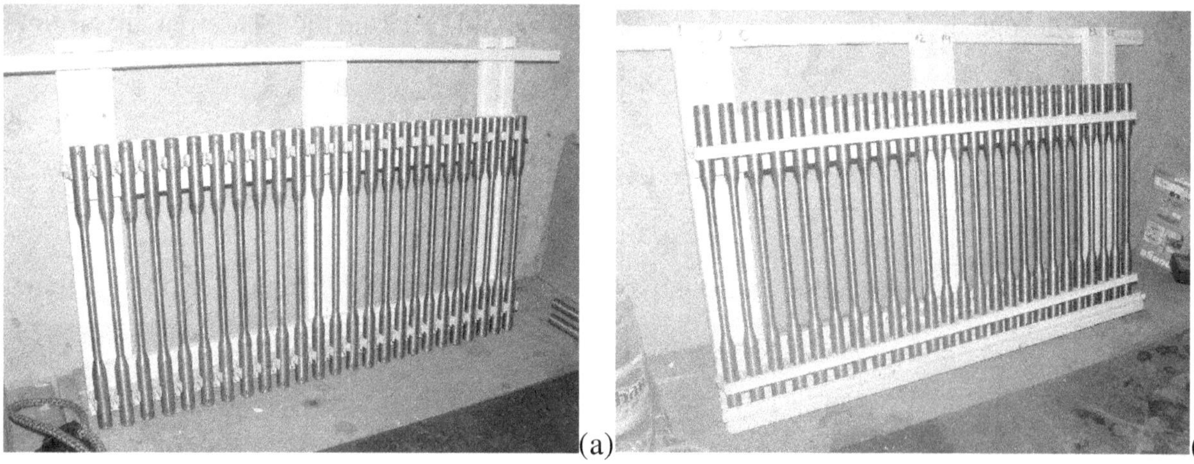

(a) (b)

Figure 3.4 Steel specimens in the corresponding matrices (a) group I; (b) group III;

Table 3.2 Number of test bodies in each group and the current used

group	I	II	III	IV	V
current [mA]	100	200	300	400	600
number of steel specimens	25	25	25	25	16

To apply the accelerated corrosion method, according to the scheme, 5 matrices with a capacity of 25 test specimens were made (Fig. 3.4 and Fig. 3.5). Each group was placed in the corresponding matrix, and then in a bath with 5% NaCl solution [49, 91] and connected to the corresponding board and/or (when necessary) the corresponding parallel connection of the stabilizers, which were previously set with the required current.

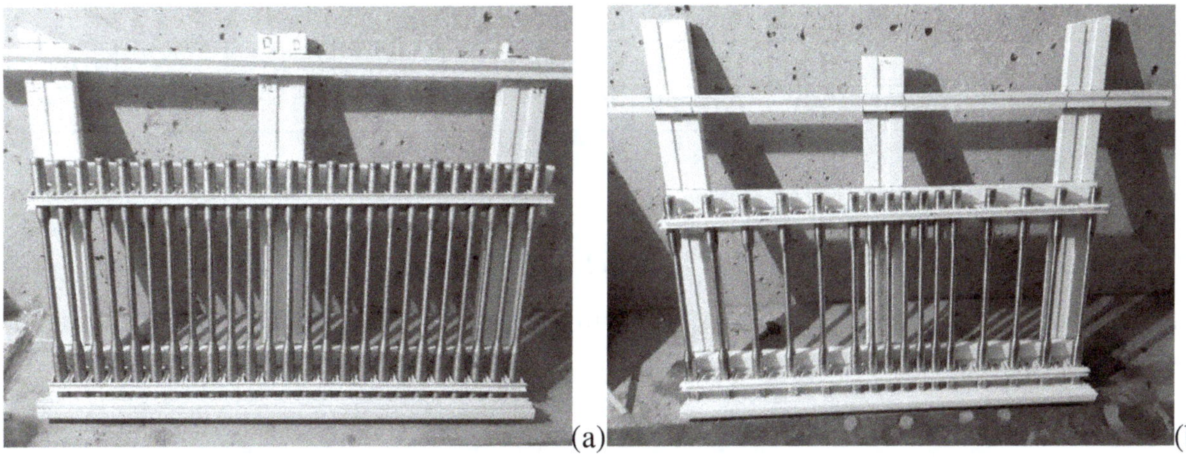

(a) (b)

Figure 3.5 Steel specimens in the corresponding matrices (a) group IV; (b) group V

Before being subjected to accelerated corrosion, the chemical composition of a test body was determined using a spectral device "Spectrotest – TXC02; S/N 4S0007; analyzer: Fe-01MO; standard sample: RH 31/12, S3-2; type of probe: SL75060798 [91], the results are given in Table 3.3. Another method for determining the elemental composition was also used, using an optical emission spectrometer "SPAS-02", the results are given in Table 3.4, moments of the determination are shown in Fig. 3.6.

Table 3.3 Chemical elements in a steel specimen [91]

Chemical elements in a steel specimen, [%]							
C	Si	Mn	P	S	Cr	Ni	Cu
0.23	0.523	1.554	0.031	0.035	0.272	0.266	0.528

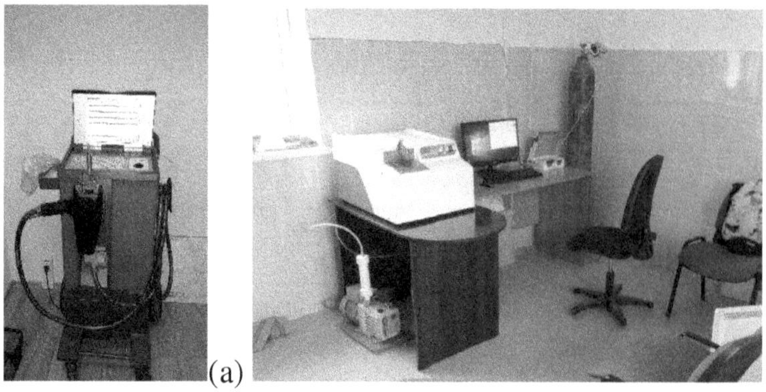

Figure 3.6 Moments of determining the chemical composition (a) With spectral device "Spectrotest – TXC02; (b) Optical emission spectrometer "SPAS-02"

Table 3.4 Chemical elements in a steel specimen [81]

Chemical elements in a steel specimen, [%]										
Fe	C	Si	Mn	P	S	Cu	N	Ni	As	Cr
96.4	0.24	0.55	1.60	0.04	0.04	0.55	0.01	0.27	0.07	0.29

2. Electrochemical accelerated corrosion

2.1. Accelerated corrosion method for steel test specimens

An electrochemical method was chosen to obtain accelerated corrosion. In terms of the stabilized quantity, the electrochemical corrosion methods are potentiostatic and galvanostatic.

In the process of material loss, the electrical resistance of the test body changes [91]. The chemical composition of the electrolyte also changes, respectively its electrical resistance [91]. It follows from Faraday's formula that the mass loss due to corrosion is directly proportional to the current flowing through the test body [91]. Therefore, if the potentiostatic method is used, periodic voltage correction will be necessary to maintain a constant rate of material loss [91]. The value of the current depends on the resistance of the steel specimen and the resistance of the electrolyte. The current through the steel specimen can be determined from Ohm's law [91] (Fig. 3.1):

$$I_{specimen} = \frac{V_{A-K}}{R_{specimen} + R_{electrolyte}} \tag{3.1}$$

Where: $I_{specimen}$ – electric current flowing through the steel specimen, $R_{specimen}$ – electrical resistance of the steel specimen, $R_{electrolyte}$ – electrical resistance of the electrolyte, V_{A-K} – voltage between anode and cathode [91].

The galvanostatic method was chosen, which avoids periodic adjustment and allows an approximate prediction of the time required to process the sample to obtain mass loss [91]. The method requires more circuit implementation than the potentiostatic method [91]. The galvanostatic method is widely used in electrochemical processes [91]. In order to ensure a stable current through each of the samples, an individual current source is required for each sample [91].

For the present study, an electrochemical corrosion system using the galvanostatic method was developed, which is also applicable to the study of other types of metal elements used in construction [91]. The system is a 25-channel adjustable current stabilizer [91]. A block diagram of the developed system is shown in Fig. 3.7 and an external view in Fig. 3.8.

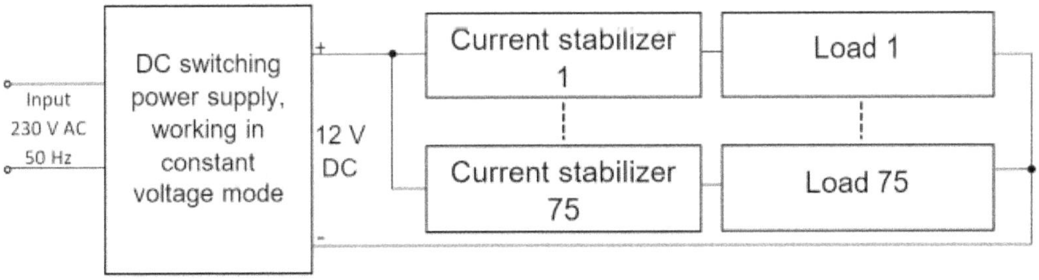

Figure 3.7 Block diagram of the developed electrochemical corrosion system [49]

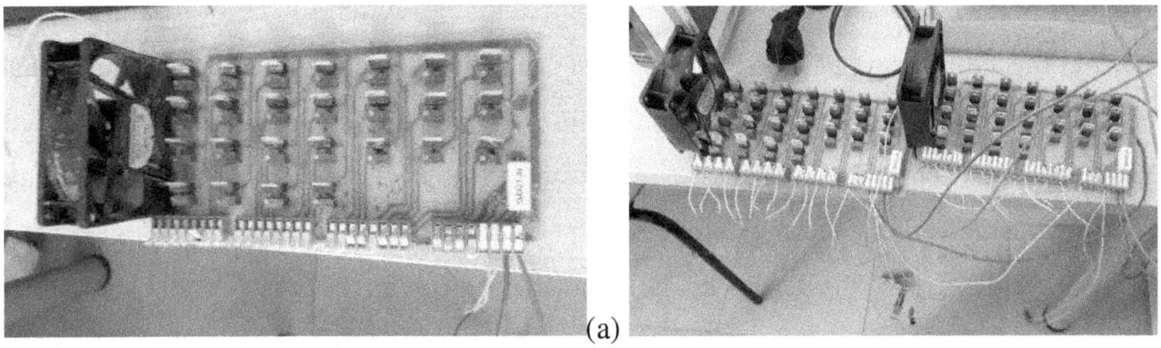

(a) (b)

Figure 3.8 (a) Single system for electrochemical corrosion by current before connection to the samples, capacity from 16 mA to 200 mA; (b) Dual system for electrochemical corrosion by current before connection to the samples, capacity from 200 mA to 400 mA

The LM317 integrated stabilizer, operating in stable current mode [91], was used to implement the individual channels. The schematic diagram of one channel is shown in Fig. 3.9 [97].

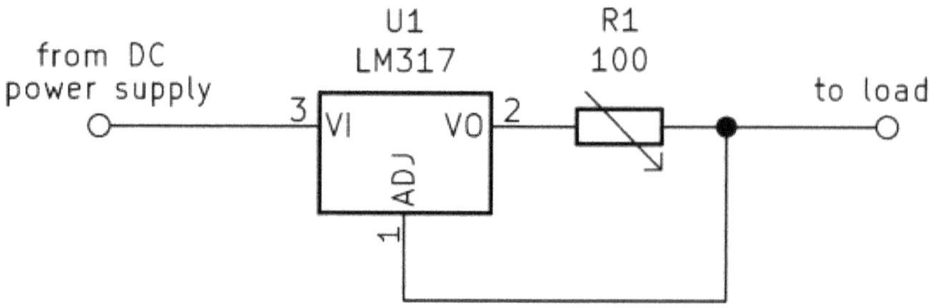

Figure 3.9 Single channel diagram [91]

The output current of each channel can be smoothly adjusted within the range from 16 mA to 200 mA at the selected value for R1 of 100 Ω [91]. The value of the output stabilized current is determined by the following formula [97] (Fig. 3.2):

$$I[\mathrm{A}] = \frac{1.25\ \mathrm{V}}{R_1} \tag{3.2}$$

The used integrated circuit LM317 has a built-in thermal protection, which limits the power dissipation [97]. This significantly increases the reliability of the developed system [91]. At the selected supply voltage of 12 V, at an output current above 100 mA, it is necessary to use additional cooling of the integrated circuit, since the junction-body thermal resistance is 50 °C/W [91]. Active cooling with a fan is used to provide additional cooling [91]. The selected circuit solution allows for parallel connection of channels. In this case, the output current is the sum of the output currents of the individual channels [91]. This means that we can choose a different current – 100 mA, 200 mA, 300 mA, 400 mA and 600 mA depending on their connection method and the settings of the generators. The percentage of material loss for one day, using Faraday's formula, for each type of different current is (Table 3.5). On fig. 3.10 shows a schematic diagram of the developed accelerated corrosion method.

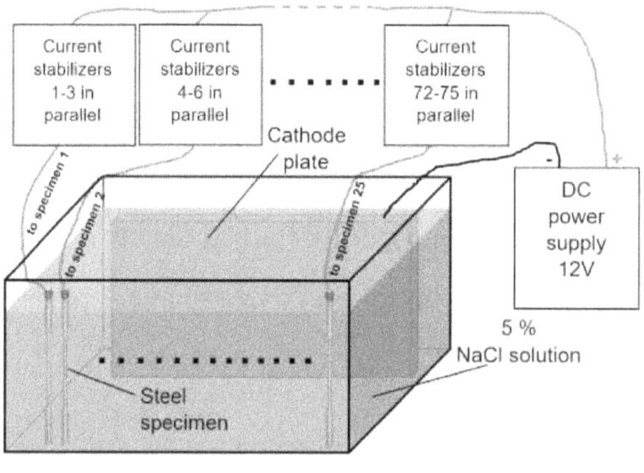

Figure 3.10 Scheme of the developed accelerated corrosion method [49]

Table 3.5 Percentage loss of material per day

current [mA]	100	200	300	400	600
weight loss – η [%]	0.54	1.07	1.62	2.14	3.21

To realize the selected current of 300 mA, each test specimen is connected to three parallel-connected channels with a certain stabilizing current for each channel of 100 mA to avoid the risk of current limitation due to the activated built-in thermal protection activated in case of overheating [91]. Possible overheating and the subsequent reduction of the current will reduce the corrosion rate for the test specimens.

2.2. Electrochemical corrosion of steel specimens

After the test specimens were distributed in matrices, they were subjected to electrochemical corrosion, according to the data in Table 3.2 – Fig. 3.11 and Fig. 3.12 (moments of electrochemical corrosion).

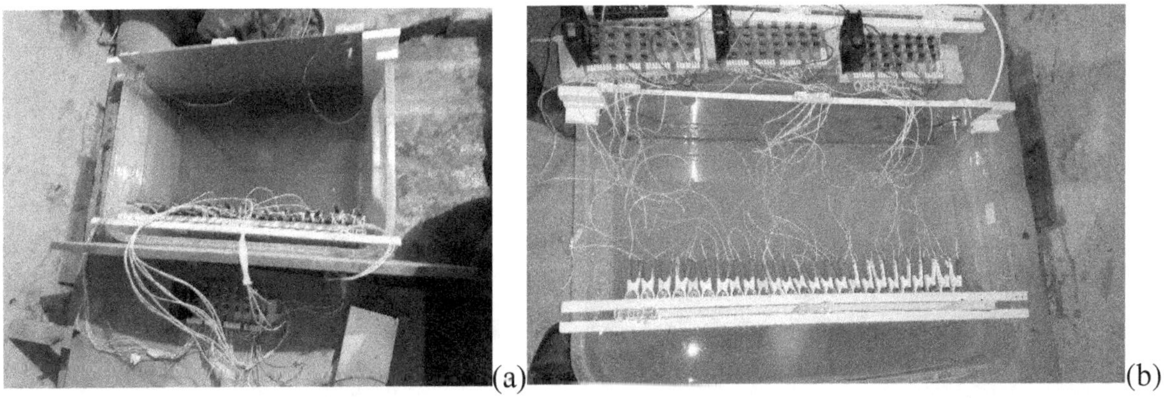

Figure 3.11 Moments of accelerated electrochemical corrosion (a) Group II [49]; (b) Group III

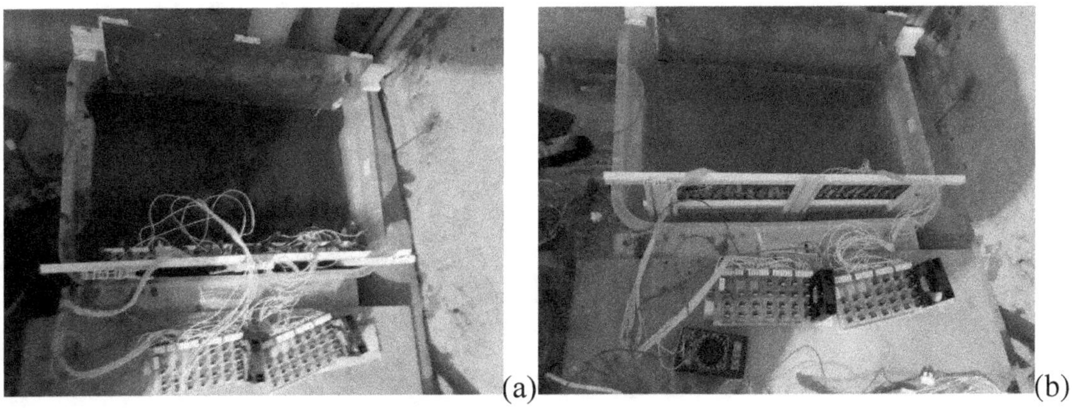

Figure 3.12 Moments of accelerated electrochemical corrosion (a) Group V; (b) Group IV

A schedule of monitoring the corrosion development is given in Fig. 3.13. The test bodies are subjected to electrochemical accelerated corrosion for a period of 72 hours. They are removed from the solution, the test body is cleaned with hydrochloric acid and the weight is measured. The test body that has reached the required mass loss is removed from the matrix and measured more accurately. The rest are placed in the bath subjected to electrochemical corrosion. Fig. 3.14 shows moments after the test bodies are removed from the bath where the electrochemical corrosion is performed. Fig. 3.15 shows test bodies before the removal of the corrosion products.

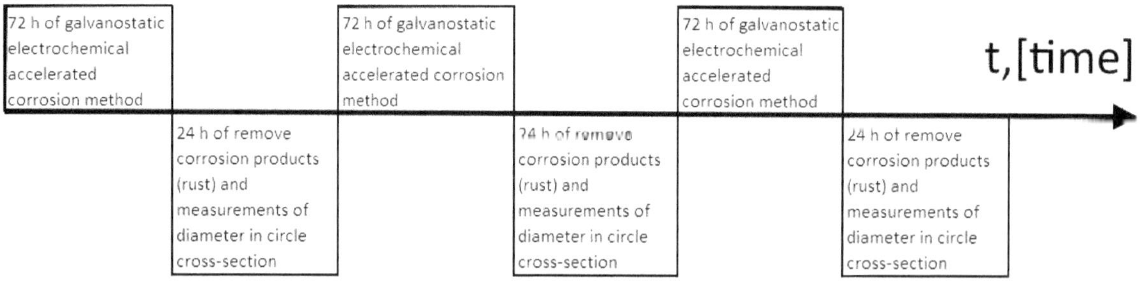

Figure 3.13 Schedule for monitoring corrosion development in steel specimens

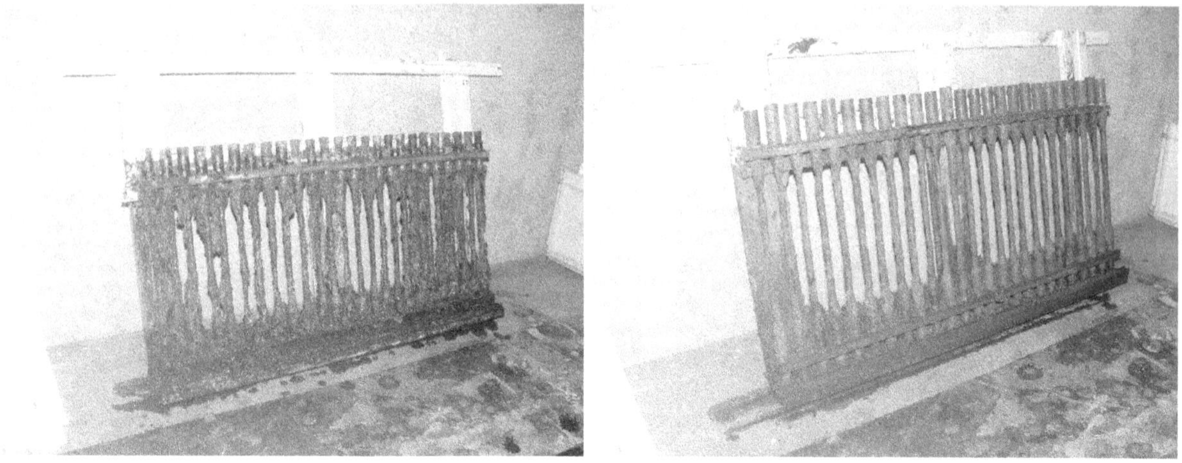

Figure 3.14 Test specimens after removal from electrochemical corrosion

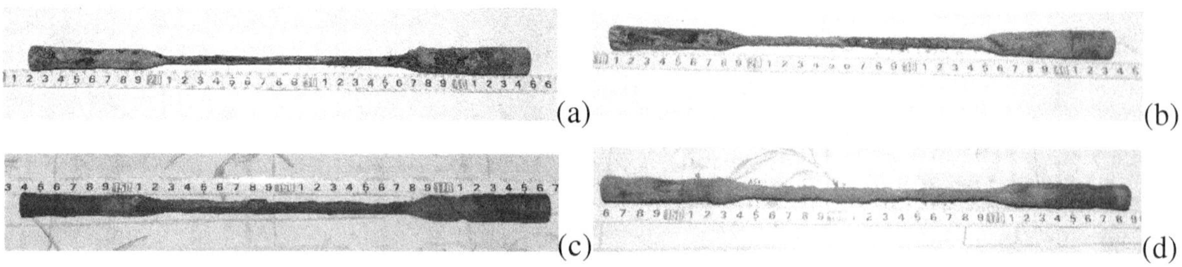

Figure 3.15 Test specimens before cleaning of corrosion products;

3. Cleaning the test bodies from corrosion products

The cleaning of the test specimens made of S355JR steel after electrochemical accelerated corrosion is carried out by a chemical method, according to the BDS ISO 8407:2014 standard, and of the methods set out in the standard, the most suitable is the method marked "C.3.5" - 10 minutes in a solution of 500 ml. HCl (hydrochloric acid), 1000 ml dH_2O (distilled water) and 3.50 grams $(CH_4)_6N_4$ (hexamethylenetetramine) [49, 81, 91]. The method is also mentioned as an applicable method in ISO 9226 (Annex A). Fig. 3.16 shows a moment of removal of corrosion products with a schematic diagram. Fig. 3.17 shows test specimens after removal of corrosion products. (хексаметилентетрамин) [49, 81, 91].

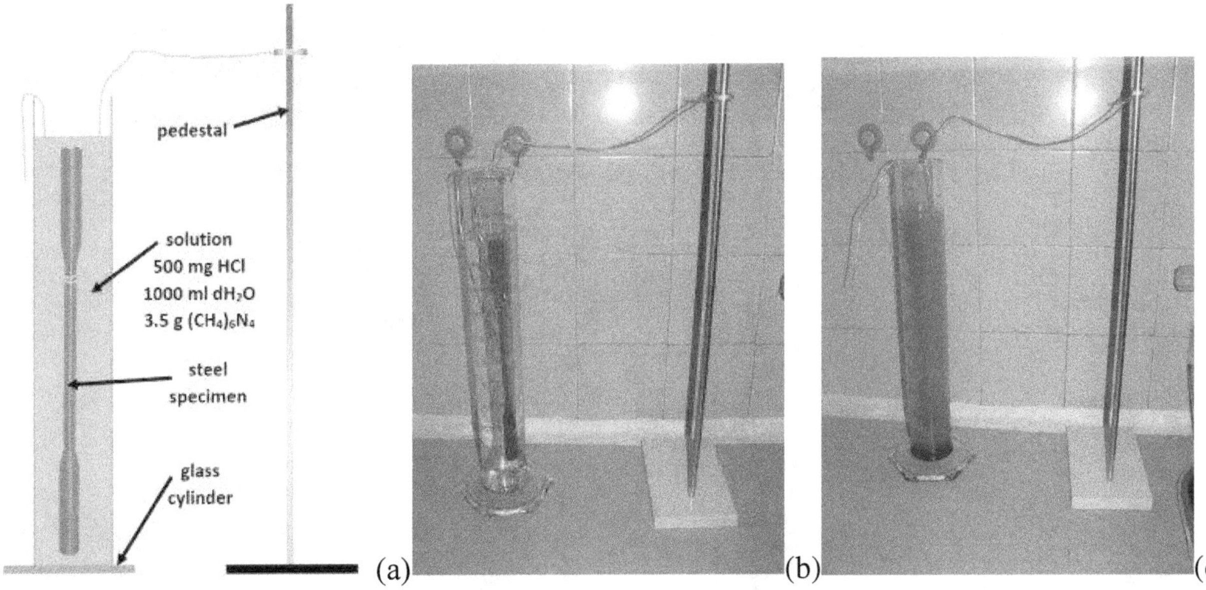

Figure 3.16 Moments of removal of corrosion products; (a) Schematic diagram for cleaning a test body [91]; (b) At the beginning [91]; (c) At the end

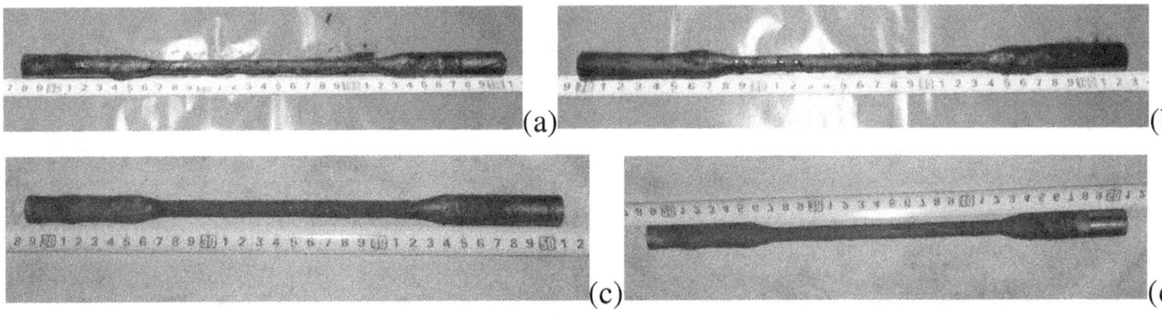

Figure 3.17 Steel test specimens after cleaning of corrosion products;

4. Measurement after electrochemical accelerated corrosion

Once it is determined that the given test body has reached its degree of corrosion, its weight is measured (to determine the loss of mass) and its diameter in the parallel length (to determine the loss of cross-section). Fig. 3.18 shows moments of weight measurement, and Fig. 3.19 shows moments of diameter measurement.

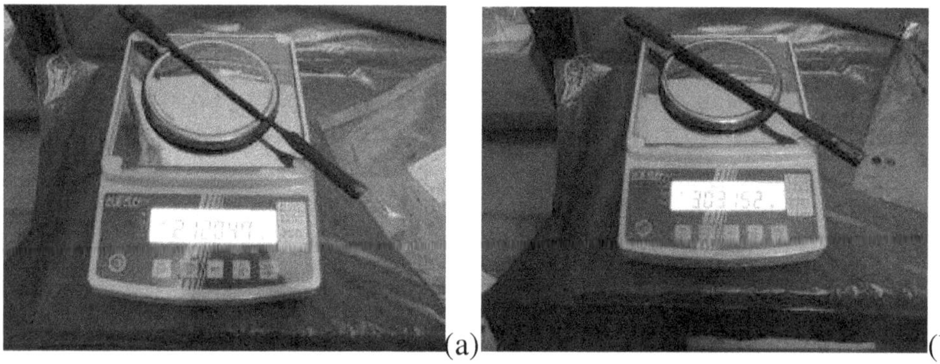
(a) (b)

Figure 3.18 Measurement of test specimens after accelerated corrosion;

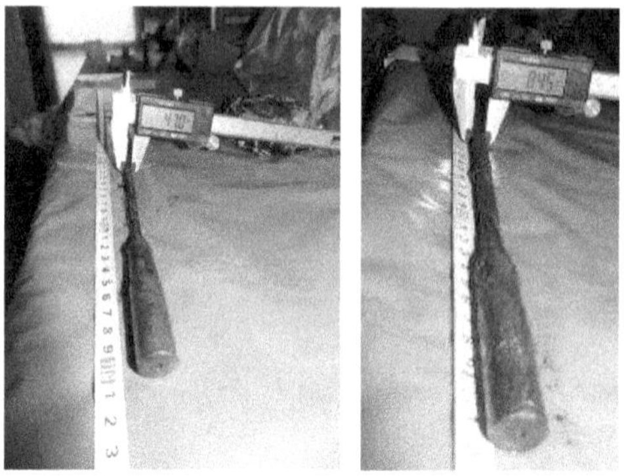

Figure 3.19 Measuring the diameter of corroded test specimens

During the measurement, it was found that the mass loss was not uniform, and the cross-section loss was also uneven. For this reason, additional measurements were made at a step (dx) of 0.50 cm in the section of the parallel length of the test body (Fig. 3.20) to determine how the cross-section had changed.

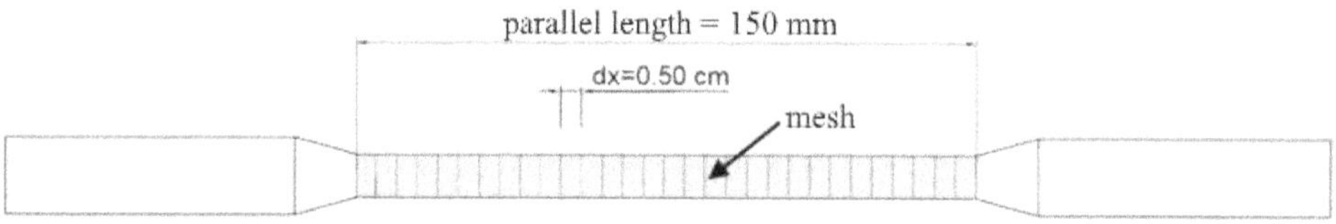

Figure 3.20 Method of measuring the change in diameter in parallel length

5. Processing of experimental data

The production technology of structural steel determines its mechanical properties, chemical composition, strength parameters, and others [89]. The quantities have a distribution that coincides with the law of normal probability distribution [89]. From the numerical axis of mechanical properties, a value is selected that has a reliability of at least 95% and is fixed as the corresponding strength characteristic of the steel, and by dividing it by a safety factor depending on the regulatory requirements, its calculated values are established [89]. With the appearance of corrosion on steel, its immediate impact begins, which is generally expressed in: a decrease in geometric characteristics, a change in strength characteristics, surface defects and structural changes [89]. When performing experiments, test specimens are used that meet the theoretical and regulatory requirements [89]. One of these requirements is the uniformity of the material and the absence of surface defects, which requirements for test specimens with corrosion cannot be practically met [89]. As a result, empirical data are obtained for which a mathematical model should be applied for processing and obtaining a final result, which result should be of sufficient reliability and be used in research work, and should have practical significance [89]. When choosing a data processing method, a connection should be sought with mathematical probability models in order to ensure practical reliability of the obtained results [89]. From the above, two main directions can be classified – stochastic and arithmetic mean.

III.B. Experimental study of mechanical properties

1. Study of steel specimens without corrosion

In order to determine the influence of corrosion on the mechanical properties of steel, a comparison should be made between results obtained as a result of corrosion and those with non-corroded material. For this purpose, eight steel specimens without corrosion (Fig. 3.22) were tested in tension according to the BDS EN ISO 6892-1:2016 standard using a MESSPHYSIK model BETA200-7/6x14 tensile testing machine. After the working diagrams of the material (Fig. 3.23) were obtained, they were processed using the stochastic method and the arithmetic mean method, the final (idealized) diagram was obtained, the values of which will serve as the basis for the comparison. In order to be able to process the results of the obtained working diagrams of the tested test specimens in practice, the stochastic method is applied as one method and the arithmetic mean method as a second method for processing the results.

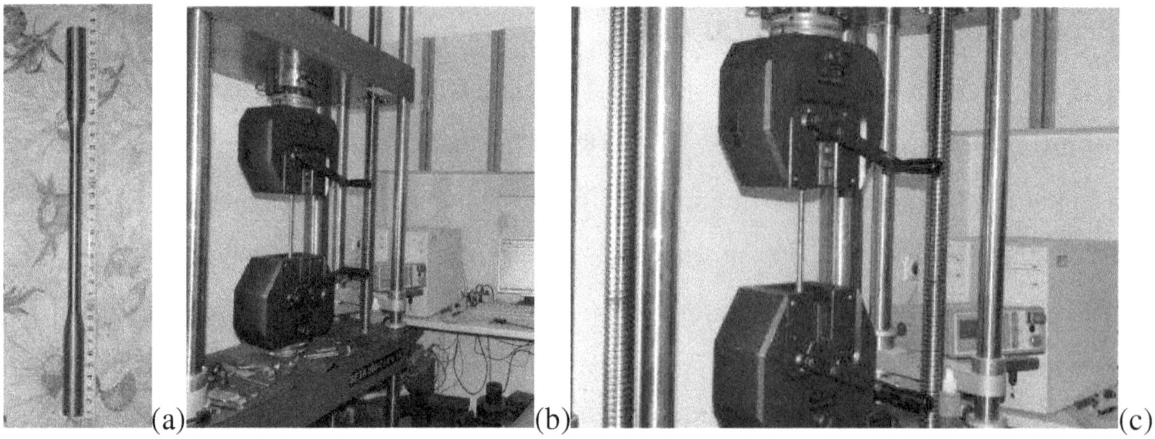

Figure 3.22 (a) Steel specimen without corrosion [81]; (b) Moments from the tensile test; (c) Moment from the tensile test;

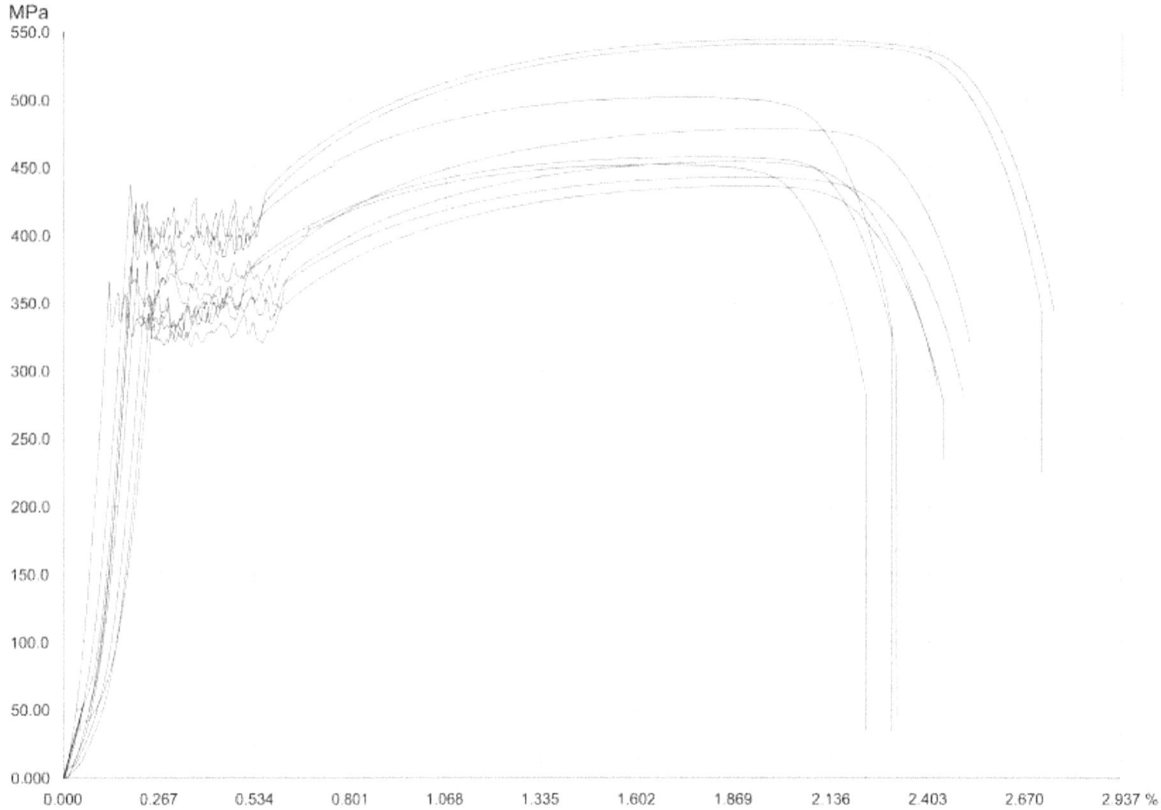

Figure 3.23 Stress-strain diagrams of S355JR steel without corrosion – 8 pieces

The stress-strain diagram of the material is divided into four main zones – elastic zone, yield zone, hardening zone and failure zone. Each of these zones is divided into eight equal parts and the values are taken – of the corresponding stress with the corresponding deformation (Fig. 3.24)

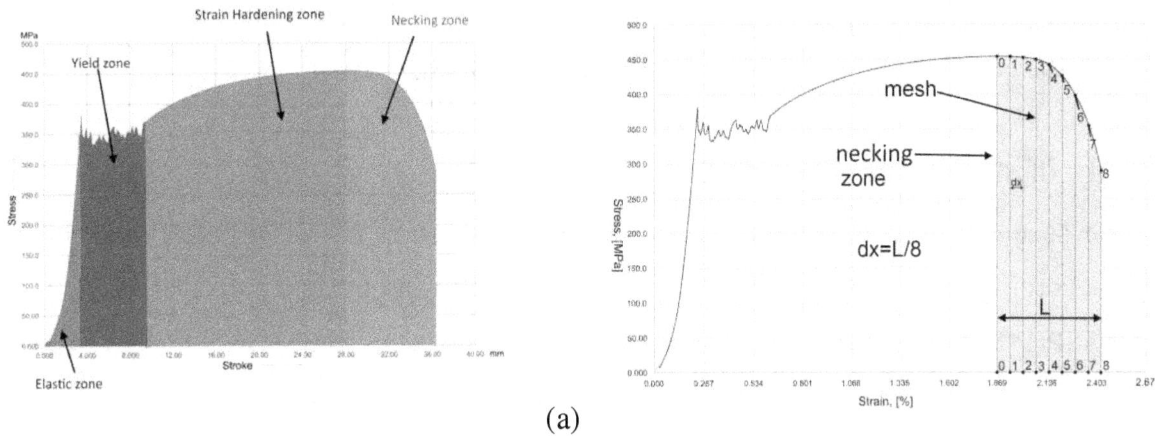

(a) (b)

Figure 3.24 How to process stress-strain curve; (a) Types of zones; (b) Zone division [49]

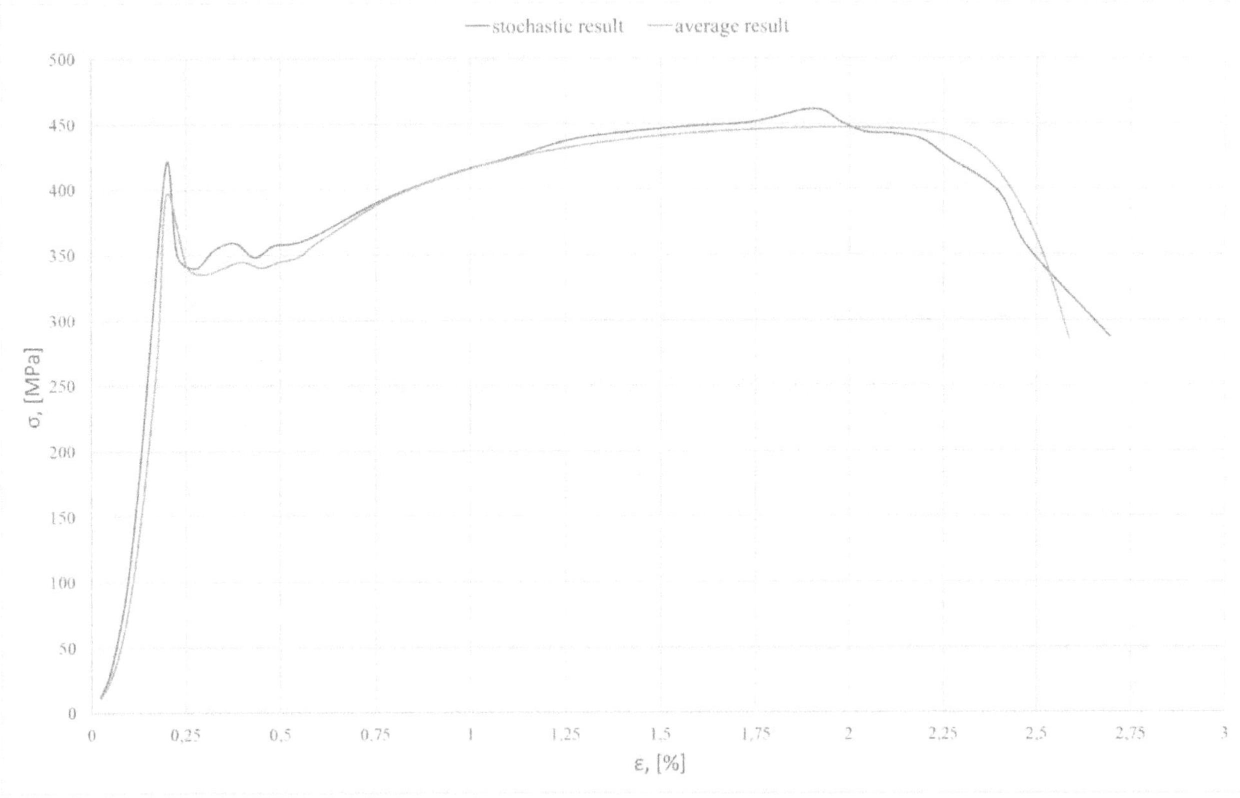

Figure 3.25 Performance stress-strain diagram of S355JR steel, after processing using the stochastic and arithmetic mean methods, which will serve as the basis for the comparison.

The processing is necessary due to the fact that the working diagrams of steel S355JR have different values of stresses and strains. The test specimens were made by applying mechanical processing, and not cast directly into a matrix of hot steel alloy. There are certain residual mechanical stresses as a result of the mechanical processing of the test specimens with the relevant machines. After performing the relevant processing, the following stress-strain diagrams of steel S355JR are obtained, which will serve as the basis for the comparison with the relevant values (Fig. 3.25).

2. Study of steel specimens with corrosion

After the specimen was measured after accelerated corrosion, it was tensile tested at room temperature according to the standard BDS EN ISO 6892-1:2016 and using a universal tensile testing machine MESSPHYSIK model BETA200-7/6x14. The tensile tests are shown in Fig. 3.26. After the specimens with corrosion were tensile tested at room temperature, the working diagrams were obtained. Processed by applying the stochastic and arithmetic mean processing method [49, 91]. Stress-strain diagrams of corroded steel are shown in Fig. 3.27.

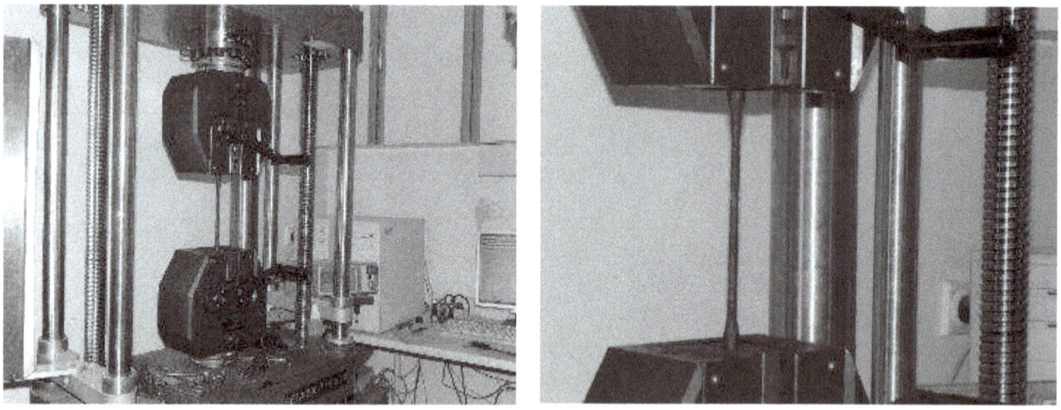

Figure 3.26 Moments from tensile testing of steel specimens with corrosion

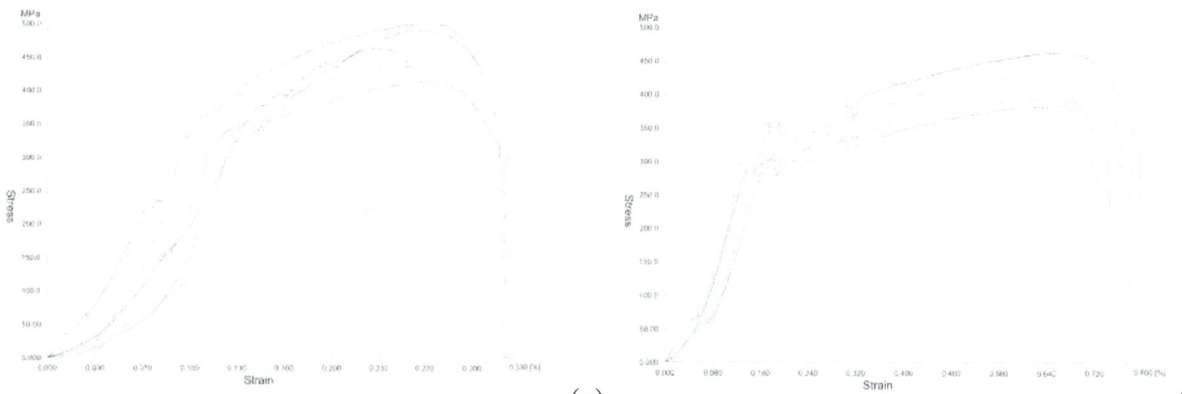

(a) (b)

Figure 3.27 Stress-strain diagrams of steel specimens with corrosion after tensile testing; (a) [91]; (b) [91];

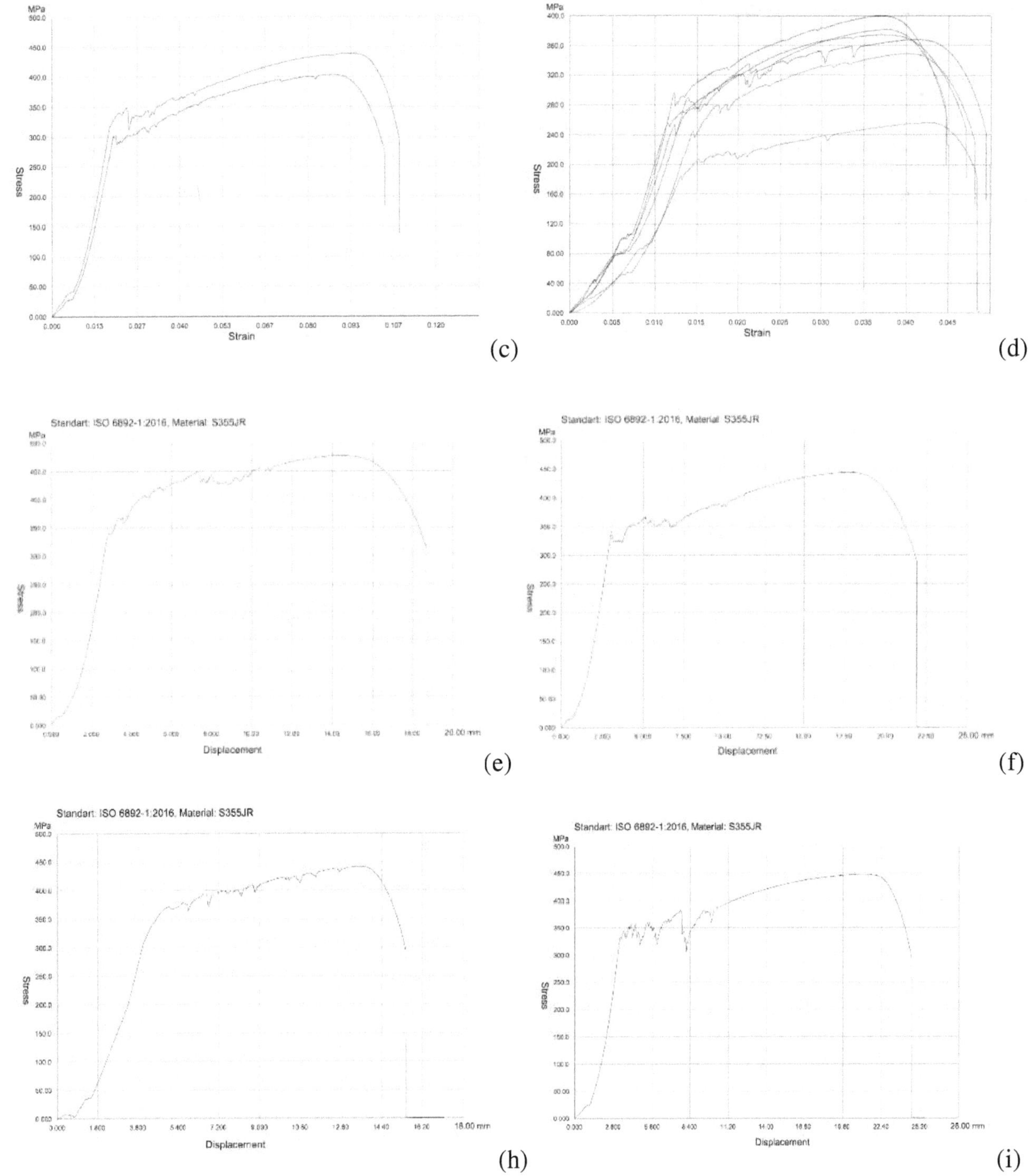

Figure 3.27 Stress-strain diagrams of steel specimens with corrosion after tensile testing; (c) [91]; (d) [91]; (e) [81]; (f) [81]; (h) [81]; (i) [81];

The processing of the obtained diagrams is carried out by processing test specimens - the diagram of the material is divided into four main zones - elastic zone, yield zone, hardening zone and failure zone.

Each of these zones is divided into eight equal parts and the values of the corresponding stress with the corresponding deformation are taken (Fig. 3.24).

2.1. Formation of groups

From the experimental results obtained, the corroded steel test specimens - 105 pieces, tested in tension at room temperature according to the standard BDS EN ISO 6892-1:2016. After processing by the stochastic and arithmetic mean method and performing an analysis of the results due to accelerated electrochemical corrosion, it is possible to group them by the change in diameter in the parallel length of the test specimen. The number of groups and the number of test specimens falling into one group are given in Table 3.6 and Fig. 3.28.

Table 3.6 Forming the groups

	group A	group B	group C	group D	group E	group F	group G	group H	group I
number	5	11	17	19	19	17	9	5	3
D [mm]	9.00 - 10.00	8.00 - 9.00	7.00 - 8.00	6.00 - 7.00	5.00 - 6.00	4.00 - 5.00	3.00 - 4.00	2.00 - 3.00	1.00 - 2.00

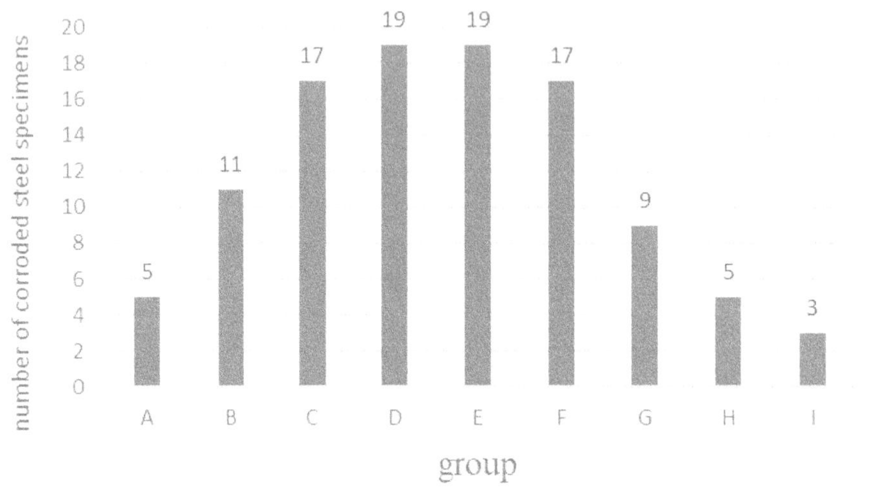

Figure 3.28 Distribution of the number of test specimens falling into the respective group, data from Table 3.6

2.2. Study of Group A

On fig. 3.29 shows the graph of the distribution of the change in diameter in the parallel length of the steel specimens group A. It consists of five corroded steel specimens with a diameter varying after the application of accelerated corrosion from 9.00 to 10.00 mm.

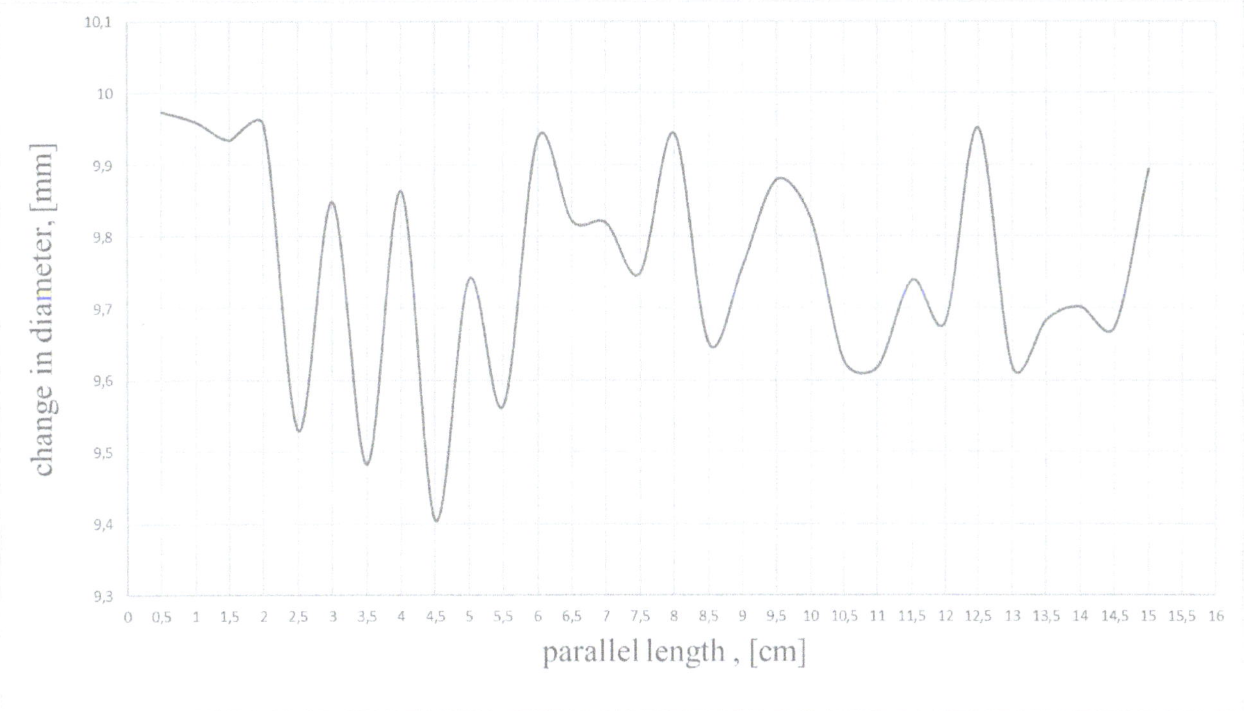

Figure 3.29 Diameter graph after accelerated corrosion in the parallel length – Group A

From Fig.3.29 it follows that the change in the diameter change due to accelerated corrosion is nonlinear and does not obey known dependencies. It corresponds most closely to the real conditions of impact, which indicates corrosion on the cross section in a corroded steel element. The application of the ISO 9223 standard gives grounds to establish that group A corresponds to corrosion impact: for corrosion category C1 - 9139 months; C2 - 475 months; C3 - 238 months; C4 - 149 months; C5 - 59 months and CX - 17 months.

2.2.1. Elastic zone

The studied zone is shown in fig.3.30 with values according to table 3.7. It is characterized that for steel without corrosion, the section is straight and follows Hooke's law. For corroded elements, the diagram begins to change smoothly, increasing the elastic deformation to 12.5% and reducing the stresses in the size to 15%. The zone becomes more elastic and less strong. The boundaries within which the band of relative deformation does not exceed 0.225%, and the band of stresses decreases and does not exceed 360 MPa.

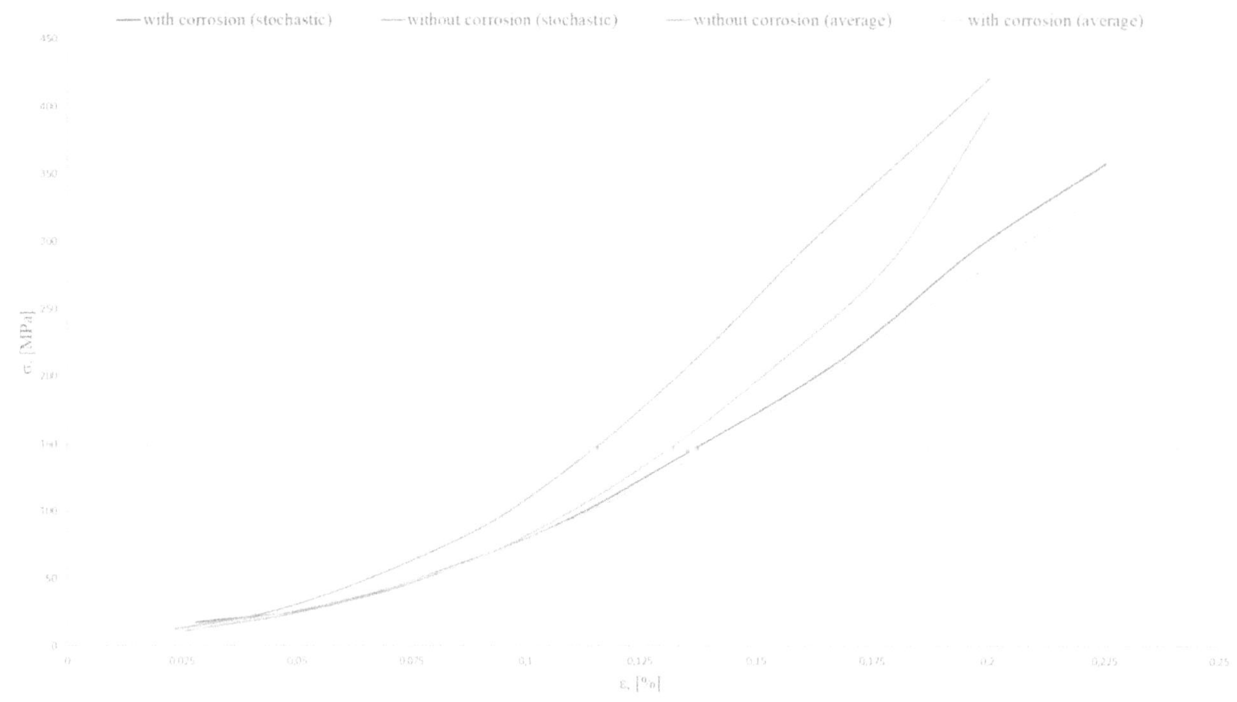

Figure 3.30 Group A elasticity zone diagrams

Table 3.7 Data on the main points on which the diagram in fig. 3.30

type of result	№ points	1	2	3	4	5	6	7	8
stochastic	ε, [%]	0.028	0.055	0.082	0.108	0.136	0.168	0.197	0.225
	σ, [MPa]	17.3	28.3	56.7	91.9	145.9	211.9	293.9	357.9
average	ε, [%]	0.028	0.057	0.085	0.113	0.141	0.170	0.198	0.226
	σ, [MPa]	12.7	31.1	59.0	98.2	149.2	211.3	277.3	338.7

2.2.2. Yield zone

In the studied section shown in fig.3.31 and table 3.8, it is established that the diagram smoothly distorts, having a clear expression in the horizontal plane. The relative deformation has not changed, i.e. the yielding area exists and is categorically definable and visible, but the stresses have decreased within the limits of up to 5%. Circumstances determine that the corroded steel is classified as a tough material and means that the destruction should occur at too large values of the deformations. There are plastic deformations in the part that do not change.

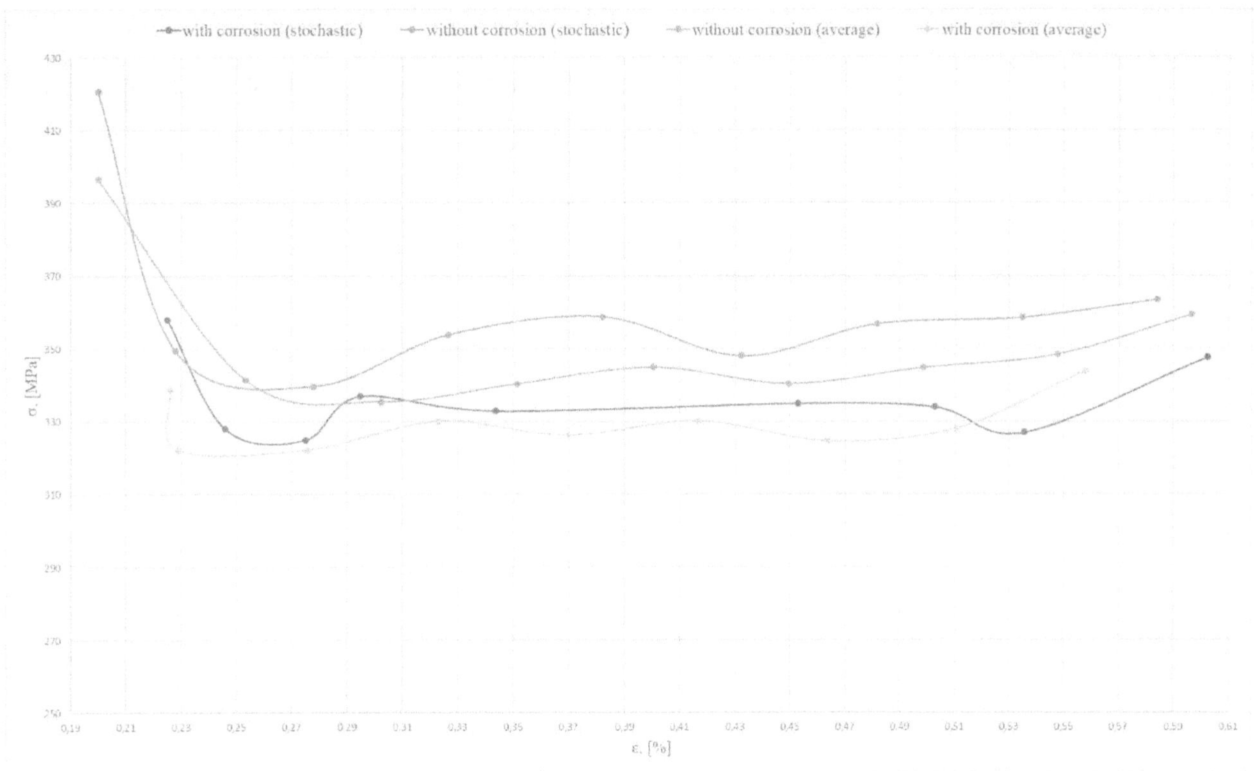

Figure 3.31 Group A yield zone diagrams

Table 3.8 Data on the main points on which the diagram in fig.3.31

type of result	№ points	1	2	3	4	5	6	7	8
stochastic	ε, [%]	0.246	0.275	0.295	0.344	0.453	0.503	0.536	0.602
	σ, [MPa]	327.9	324.9	336.9	332.9	334.9	333.9	326.9	347.4
average	ε, [%]	0.229	0.276	0.323	0.370	0.417	0.464	0.511	0.558
	σ, [MPa]	322.2	322.2	330.0	326.2	330.1	324.6	328.1	343.7

2.2.3. Strain hardening zone

In the studied area of fig.3.32 and table 3.9, it is found that the strip relative deformation has not changed, and the stresses follow the specified decrease to 5%. The nonlinear development of the mechanical properties is preserved, expressed in a smooth increase in stress and the presence of an increase in deformation. It is known that steel has a polycrystalline structure, which means that the volume in the steel contains a large number of small crystals that have a chaotic orientation. The atoms of the metal are arranged in the crystals in a certain way, building a regular spatial lattice. In the lattice there are forces of interaction between the atoms, and they form a system that is completely determined for each steel, i.e. there is no significant change in the crystal lattice, the sliding of each individual crystal cannot grow unlimitedly - the presence of a changed structure as a result of plastic deformation is the weaker slope in the diagram.

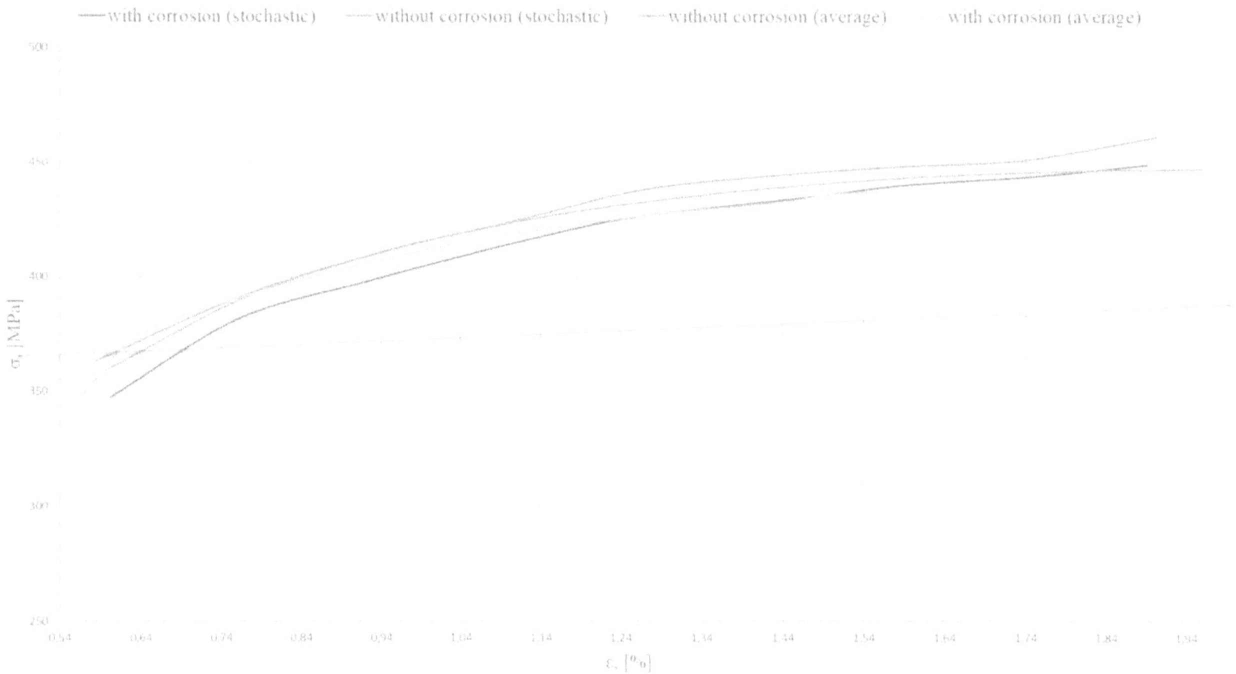

Figure 3.32 Group A strain hardening zone diagrams

Table 3.9 Data on the main points on which the diagram in fig.3.32

type of result	№ points	1	2	3	4	5	6	7	8
stochastic	ε, [%]	0.759	0.926	1.121	1.257	1.453	1.589	1.755	1.892
	σ, [MPa]	381.4	398.4	415.9	425.9	433.9	439.9	443.9	448.9
average	ε, [%]	0.638	0.799	0.959	1.119	1.280	1.440	1.494	1.560
	σ, [MPa]	374.8	394.6	409.0	419.4	427.0	432.3	435.7	437.1

2.2.4. Necking zone

In the studied area of fig.3.33 and table 3.10, it is determined that the deformation limit decreases to 12%, the fracture stress does not change. The formation of the neck is observed, i.e. relatively large deformations are localized in a small area of the material and a significant reduction in the cross-sectional area occurs. It is found that there is heterogeneity of the corroded steel. As the material begins to deform, all places undergo approximately the same amount of deformation and the greater the local stress, the greater the local reduction in the cross-sectional area, which in turn leads to an even greater concentration of stresses, there is instability, which causes the formation of the neck.

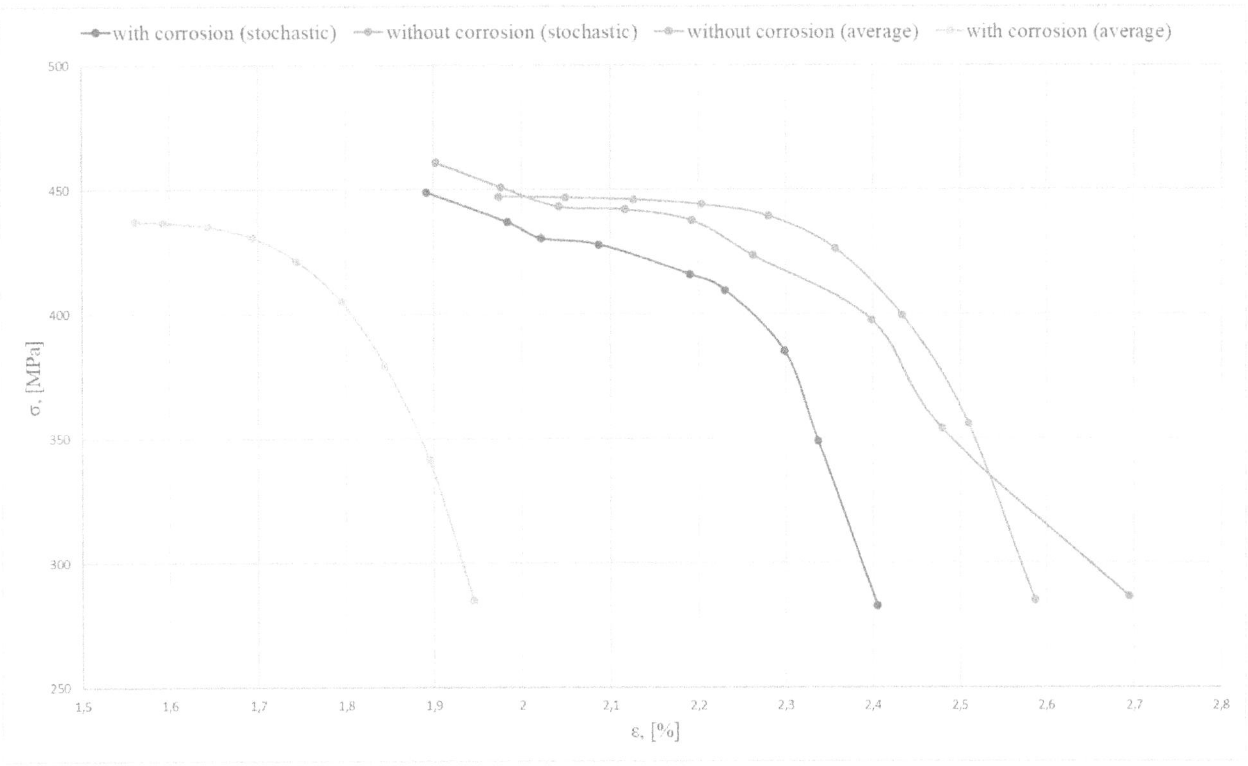

Figure 3.33 Group A necking zone diagrams

Table 3.10 Data on the main points on which the diagram in fig.3.33

type of result	№ points	1	2	3	4	5	6	7	8
Stochastic result	ε, [%]	1.983	2.022	2.087	2.190	2.230	2.298	2.337	2.405
	σ, [MPa]	436.9	430.4	427.8	415.9	409.5	385.0	348.9	282.7
Average result	ε, [%]	1.593	1.644	1.694	1.744	1.794	1.845	1.895	1.945
	σ, [MPa]	436.7	435.1	430.7	421.4	405.2	379.1	341.3	284.9

2.2.5. Stress-strain diagram

From the studied stress-strain diagram shown in fig. 3.34 of the corroded material, it can be clearly deduced that corrosion indicates its influence on the mechanical properties of the studied steel, expressed in the following basic facts:

- reduction of steel ductility by about 12%
- the structure of the material does not change significantly and it remains ductility

- pronounced yield strength
- slight reduction in strength values up to 5% – 15%
- the internal crystal structure of the steel does not change significantly
- the modulus of elasticity – decreases by about 20%

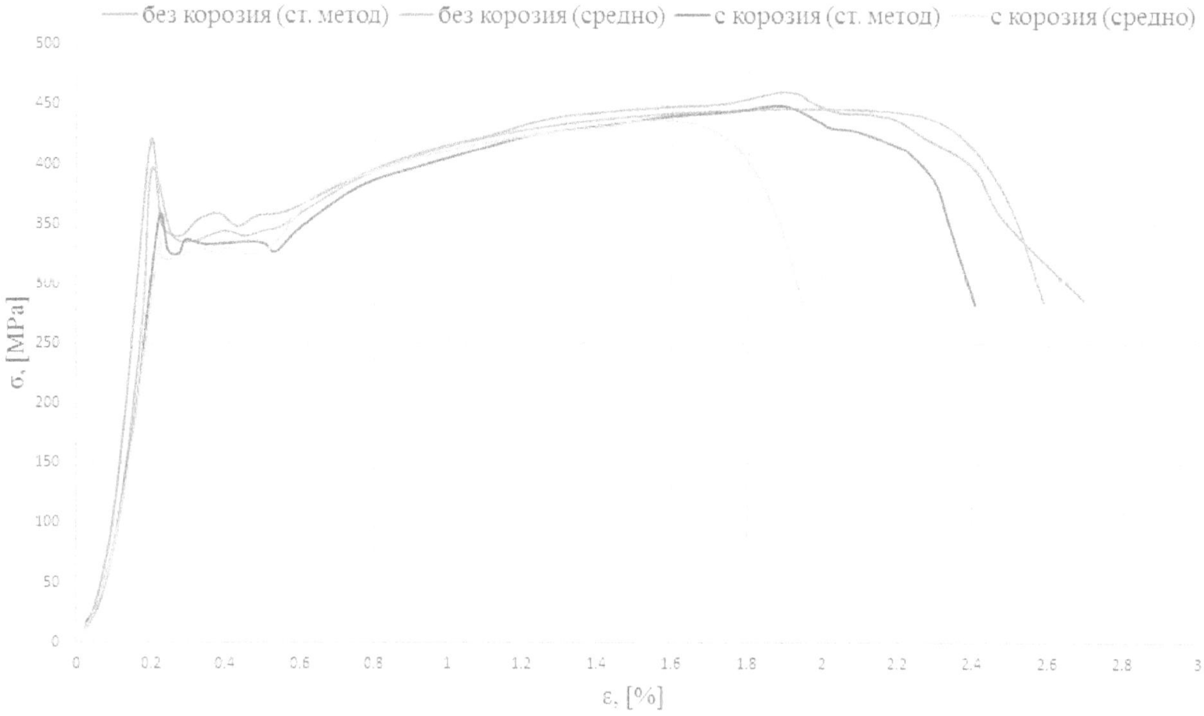

Figure 3.34 Group A stress-strain diagrams

2.3. Study of Group B

Group B consists of 11 corroded steel test specimens with a diameter varying after application of accelerated corrosion from 8.00 to 9.00 mm. It is established from Fig. 3.35 that the change in the diameter change as a result of accelerated corrosion is nonlinear and does not obey known dependencies, which fact brings us as close as possible to the real conditions of impact, which indicates corrosion on the cross section in a corroded steel element. The application of the ISO 9223 standard gives grounds to establish that group B corresponds to corrosion impact: for corrosion category C1 - 14769 months; C2 - 768 months; C3 - 384 months; C4 - 240 months; C5 - 96 months and CX - 28 months.

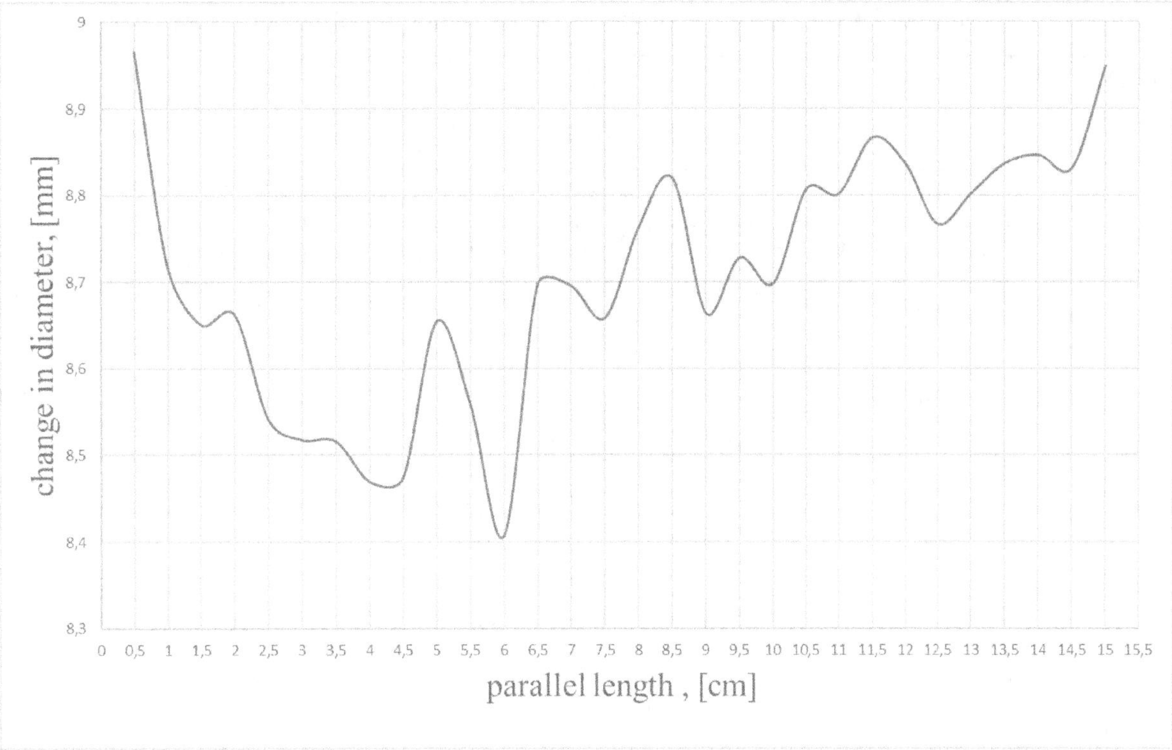

Figure 3.35 Diameter graph after accelerated corrosion in the parallel length – Group B

2.3.1. Elastic zone

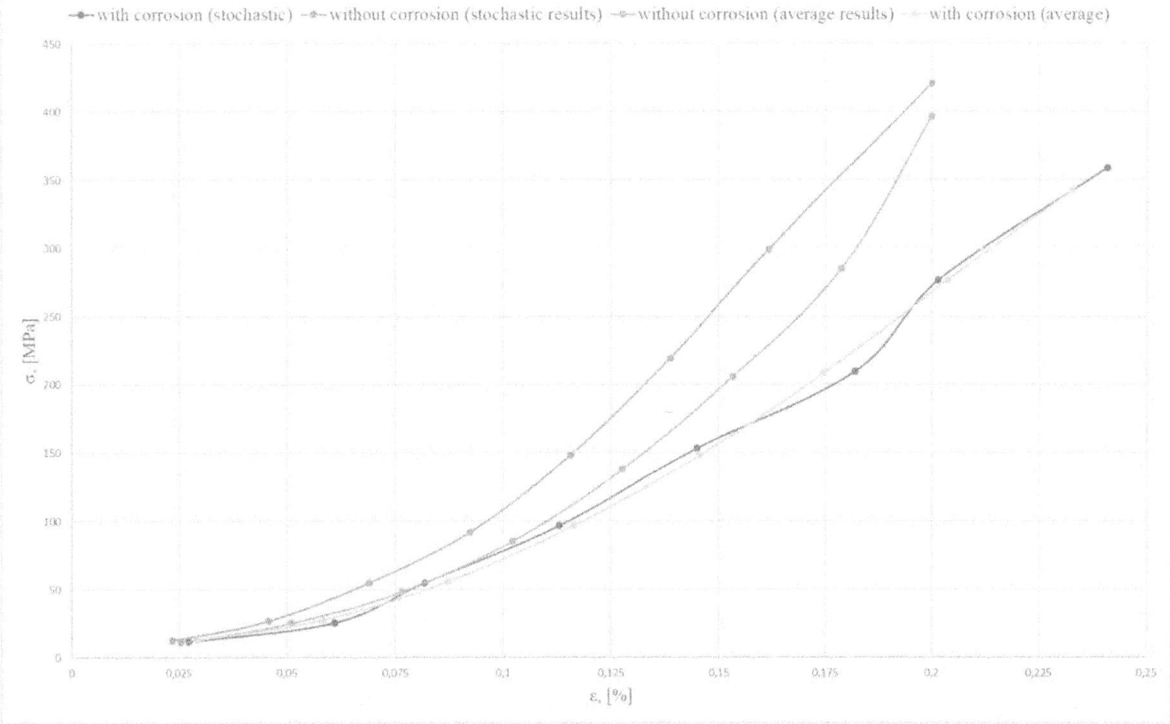

Figure 3.36 Group B elasticity zone diagrams

Table 3.11 Data on the main points on which the diagram in fig.3.36

type of result	№ points	1	2	3	4	5	6	7	8
stochastic	ε, [%]	0.027	0.061	0.082	0.113	0.145	0.182	0.202	0.241
	σ, [MPa]	11.4	25.3	54.6	96.5	153.1	209.3	276.5	358.9
average	ε, [%]	0.029	0.058	0.087	0.116	0.145	0.175	0.204	0.233
	σ, [MPa]	12.8	27.4	55.9	96.7	147.9	208.4	276.6	342.6

The studied segment in fig.3.36 and table 3.11 is characterized by the fact that for steel without corrosion, the section is straight and follows Hooke's law. For corroded elements, it was found that the diagram begins to change, increasing the elastic deformation to 20% and leading to a decrease in the stresses in the size to 15%. In general, the zone becomes more elastic and less strong. The boundaries in which the strip is located, the relative deformation does not exceed 0.24%, and the strip stresses decrease and do not exceed 360 MPa. Signs of nonlinear behavior of the elastic modulus begin to appear, as a whole it decreases.

2.3.2. Yield zone

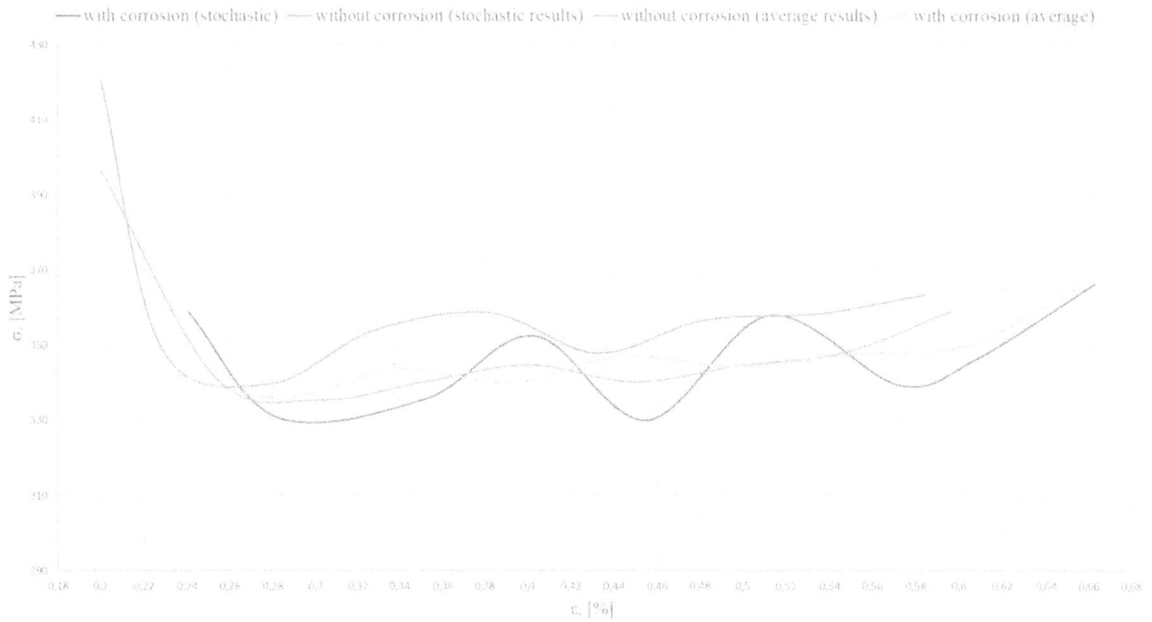

Figure 3.37 Group B yield zone diagrams

In the studied area shown in fig.3.37 and table 3.12, it is established that the diagram is smoothly distorted, with a clear expression in the horizontal plane. It can be assumed that the relative deformation has not changed, i.e. the yield point exists and is categorically definable, but the stresses have a slight decrease in the range of up to 5%. Corroded steel is classified as a tough material and means that the failure should occur at very large values of deformations. There are plastic deformations that do not change compared to non-corroded steel.

Table 3.12 Data on the main points on which the diagram in fig.3.37

type of result	№ points	1	2	3	4	5	6	7	8
stochastic	ε, [%]	0.283	0.351	0.403	0.455	0.512	0.572	0.608	0.663
	σ, [MPa]	330.6	335.4	352.4	329.9	357.9	339.4	346.4	366.4
average	ε, [%]	0.286	0.339	0.391	0.444	0.497	0.550	0.603	0.656
	σ, [MPa]	336.1	343.9	340.1	347.1	344.3	347.7	349.9	365.9

2.3.3. Strain hardening zone

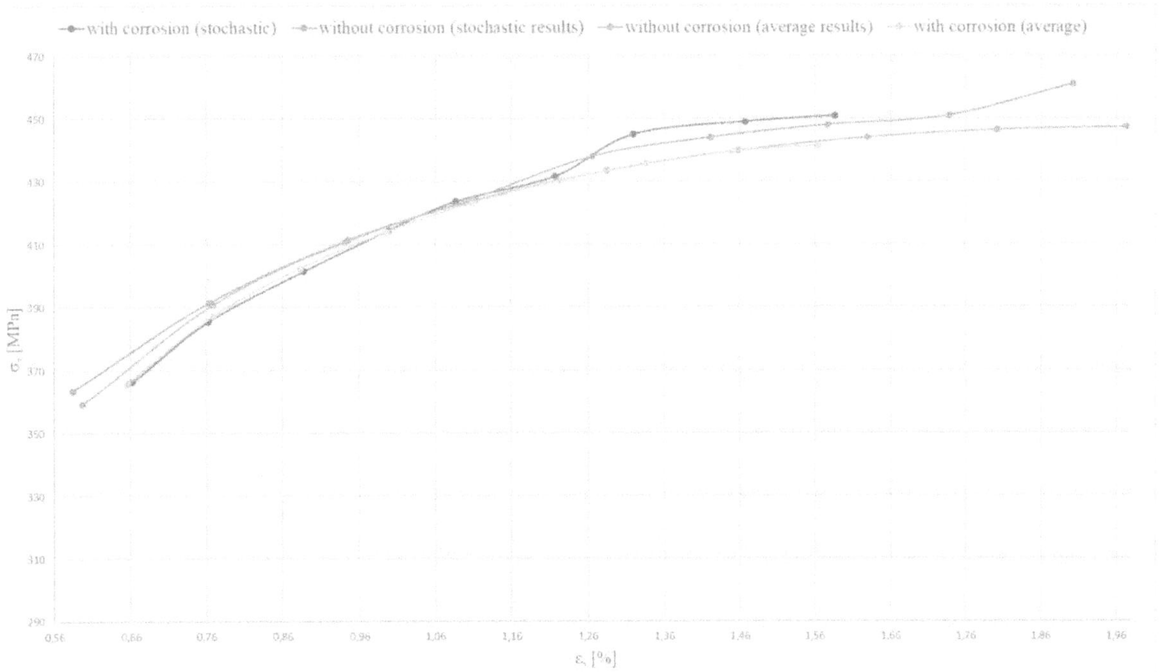

Figure 3.38 Group B strain hardening zone diagrams

In the studied part of fig.3.38 and table 3.13, it is found that the strip of relative deformation begins a change expressed in a decrease of up to 20%, and the stresses follow a decrease of up to 5%. The nonlinear development of mechanical properties is preserved, expressed in a smooth increase in stress and the presence of an increase in deformation. It is known that steel has a polycrystalline structure, which means that the volume in the steel contains a large number of small crystals that have a chaotic and random orientation. The atoms of the metal are arranged in the crystals in a certain way, building a regular spatial lattice. In a lattice there are forces of interaction between the atoms, they form a system that is completely defined for each steel. A change in the crystal lattice begins.

Table 3.13 Data on the main points on which the diagram in fig.3.38

type of result	№ points	1	2	3	4	5	6	7	8
stochastic	ε, [%]	0.762	0.888	0.999	1.087	1.216	1.320	1.467	1.586
	σ, [MPa]	385.4	401.4	414.5	423.6	431.6	444.9	448.9	450.9
average	ε, [%]	0.769	0.883	0.996	1.109	1.223	1.336	1.450	1.563
	σ, [MPa]	387.2	402.2	413.9	423.1	430.2	435.7	439.5	441.3

2.3.4. Necking zone

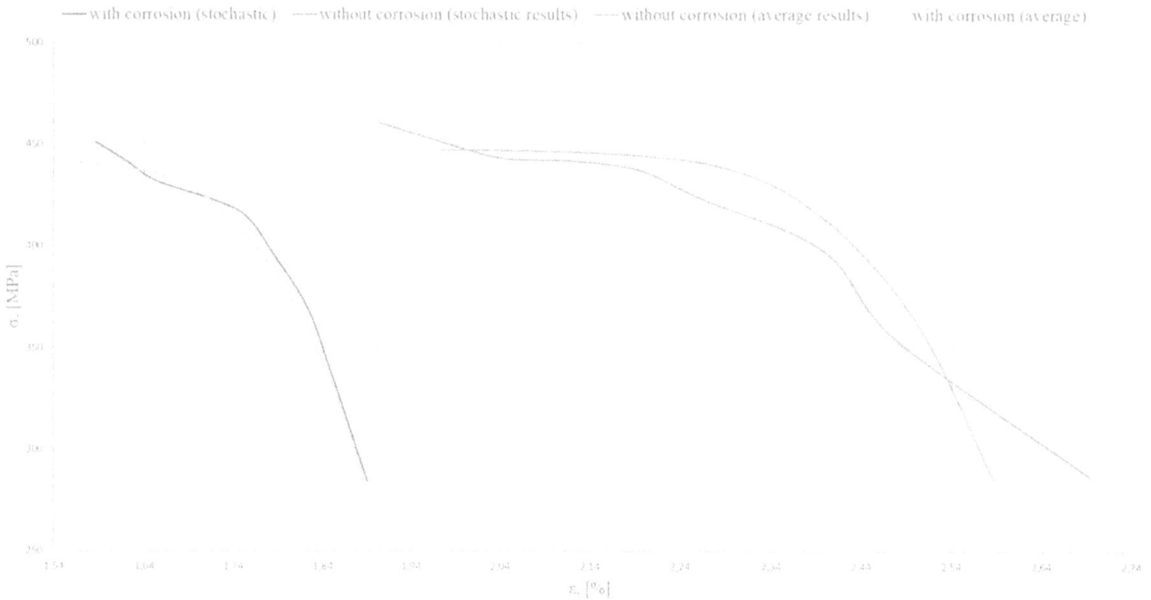

Figure 3.39 Group B necking zone diagrams

From fig. 3.39 and table 3.14, it is established that the deformation limit decreases to 30%, the failure stress does not change. The formation of the neck is observed, i.e. relatively large deformations are localized in a small part of the material and a significant reduction in the cross-sectional area occurs. It can be assumed that there is heterogeneity of the corroded steel. As the material begins to deform, all places undergo approximately the same amount of deformation. The greater the local stress, the greater the local reduction in the cross-sectional area, which in turn leads to an even greater concentration of stresses. It reaches instability, which causes the formation of the neck. Failure occurs at deformation values characteristic of the tensile strength of uncorroded steel.

Table 3.14 Data on the main points on which the diagram in fig.3.39

type of result	№ points	1	2	3	4	5	6	7	8
stochastic	ε, [%]	1.621	1.654	1.716	1.754	1.785	1.824	1.851	1.891
	σ, [MPa]	441.4	431.8	422.8	414.5	395.6	368.9	335.9	284.4
average	ε, [%]	1.603	1.643	1.683	1.723	1.763	1.803	1.843	1.884
	σ, [MPa]	439.4	436.6	431.7	419.3	400.3	374.3	339.5	290.5

2.3.5. Stress-strain diagram

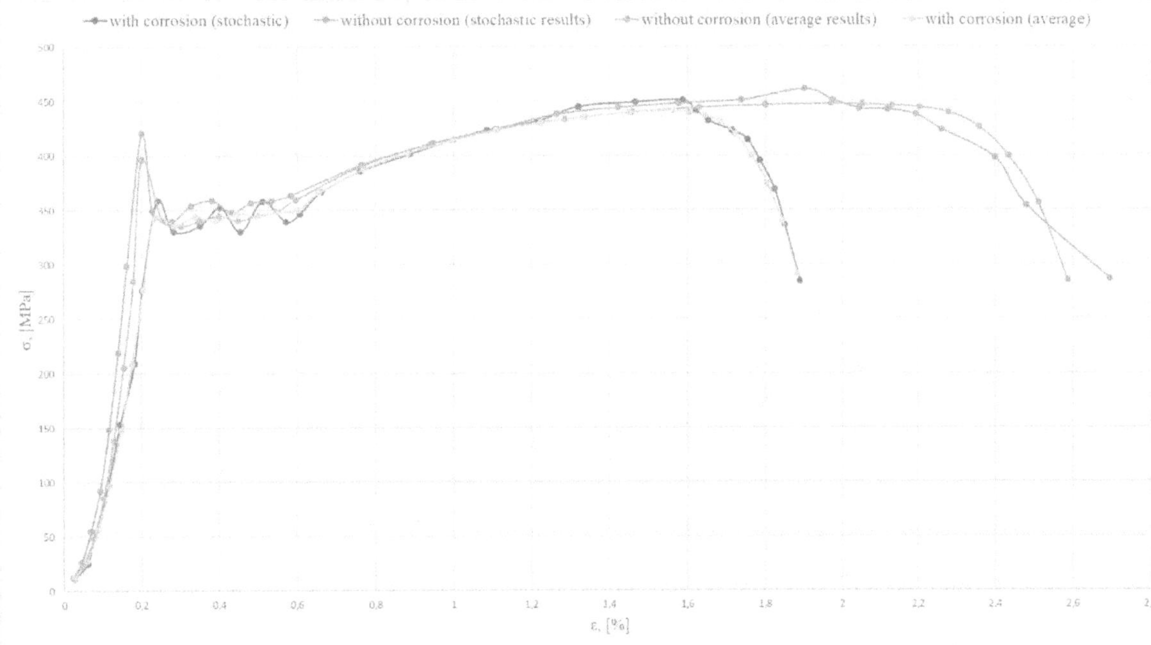

Figure 3.40 Group B stress-strain diagrams

From the studied stress-strain diagram in fig. 3.40 of the corroded material, it can be clearly deduced that corrosion has its influence on the mechanical properties of the studied steel, expressed in the following basic facts:

- reduction of steel ductility by about 30%
- there is a slight change in the structure of the material, but it continues to remain a ductility material.

- a pronounced yield strength
- slight reduction in strength values up to 5% – 15%
- signs of changes in the internal crystal structure of the steel begins
- the modulus of elasticity – decreases by about 20%

2.4. Study of Group C

Group C consists of 17 corroded steel test specimens with a diameter ranging from 7.00 to 8.00 after application of accelerated corrosion. It is established from Fig. 3.41 that the change in the diameter change as a result of accelerated corrosion is nonlinear. It does not obey known dependencies, which fact brings us as close as possible to the real conditions of impact, which indicates corrosion on the cross-section in the steel element. The application of the ISO 9223 standard gives grounds to establish that group B corresponds to corrosion impact: for corrosion category C1 - 24923 months; C2 - 1296 months; C3 - 648 months; C4 - 405 months; C5 - 162 months and CX - 46 months.

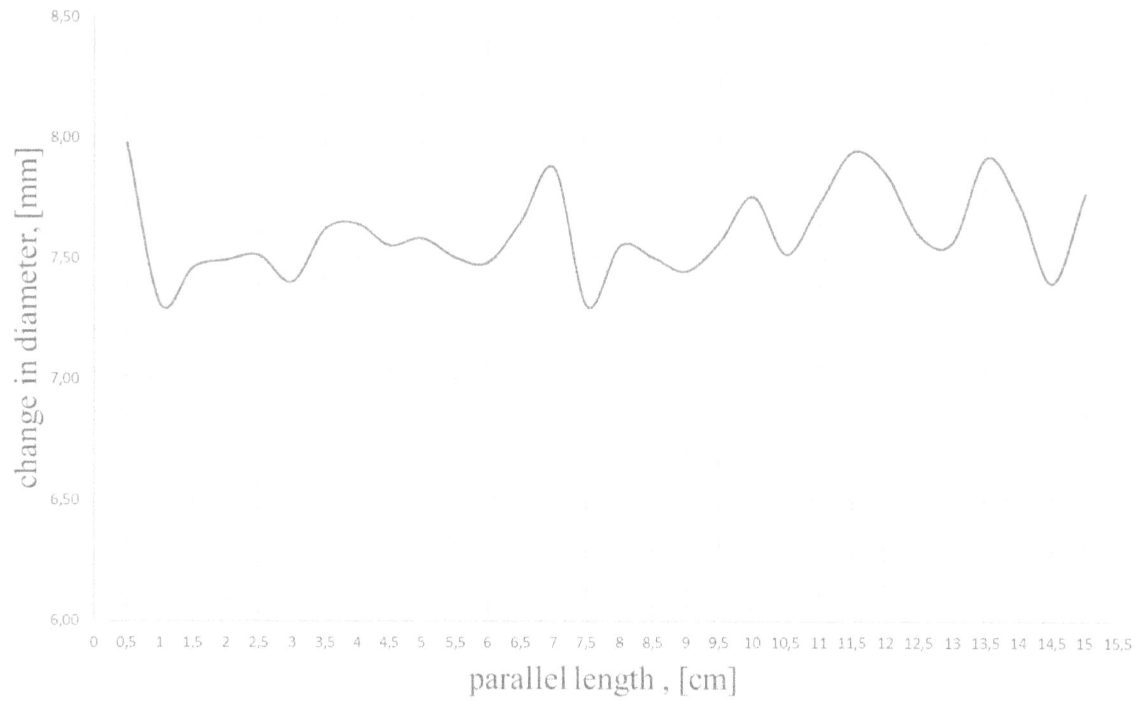

Figure 3.41 Diameter graph after accelerated corrosion in the parallel length – Group C

2.4.1. Elastic zone

From fig.3.42 and table 3.15, it is characterized that for steel without corrosion, the section is straight and follows Hooke's law. For corroded elements, it is found that the diagram begins to change, increasing the elastic deformation to 20% and leading to a 15% reduction in stresses. The area becomes more elastic and less strong. The boundaries within which the strip of relative deformation does not exceed 0.24%. There is evidence of nonlinear behavior of the modulus of elasticity, which decreases.

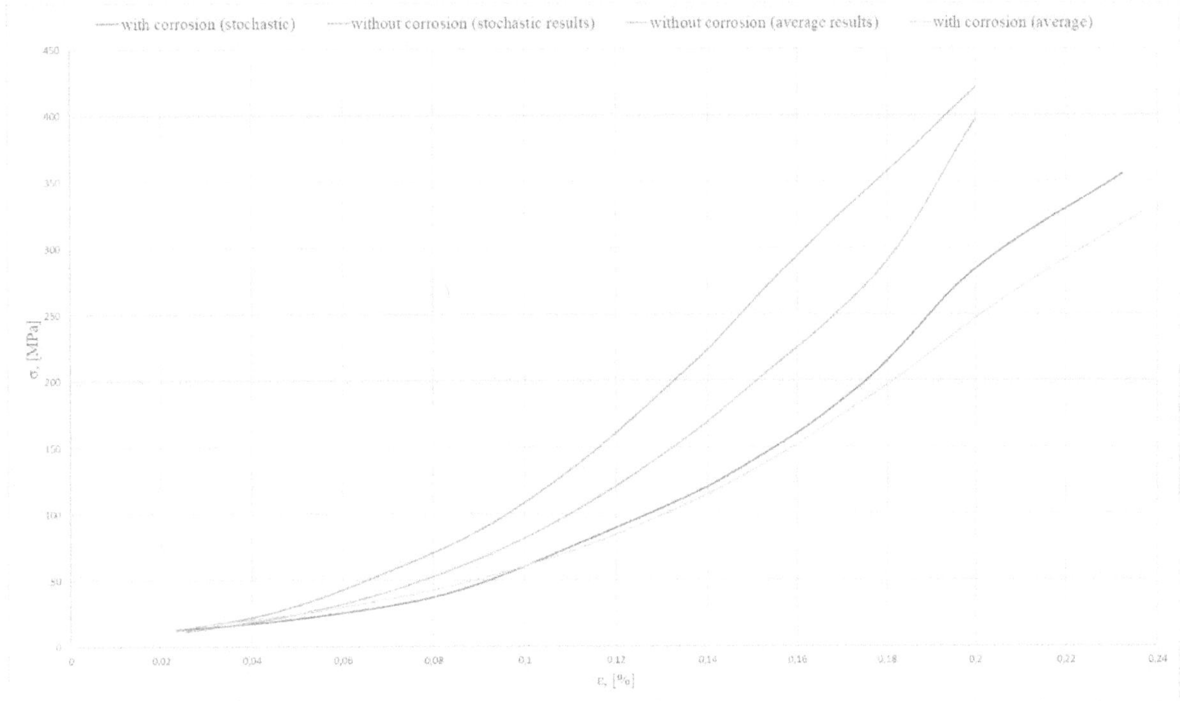

Figure 3.42 Group C elasticity zone diagrams

Table 3.15 Data on the main points on which the diagram in fig.3.42

type of result	№ points	1	2	3	4	5	6	7	8
stochastic	ε, [%]	0.023	0.053	0.084	0.113	0.143	0.173	0.200	0.232
	σ, [MPa]	12.3	22.2	40.5	79.6	125.6	193.0	283.5	355.6
average	ε, [%]	0.030	0.059	0.089	0.118	0.148	0.178	0.207	0.237
	σ, [MPa]	15.8	29.3	49.7	82.0	128.4	191.3	264.1	327.0

2.4.2. Yield zone

In the studied section of fig.3.43 and table 3.16, it is established that the diagram has a clear expression in the horizontal plane, but with a steep effect in its vertical direction. It can be assumed that the relative deformation has not changed, i.e. the yield point exists and is categorically definable, but the stresses have a slight decrease in the range of up to 5%. Corroded steel can still be classified as a tough material and means that the failure should occur at very large values of the deformations. There are plastic deformations in the part, which do not change compared to the non-corroded steel.

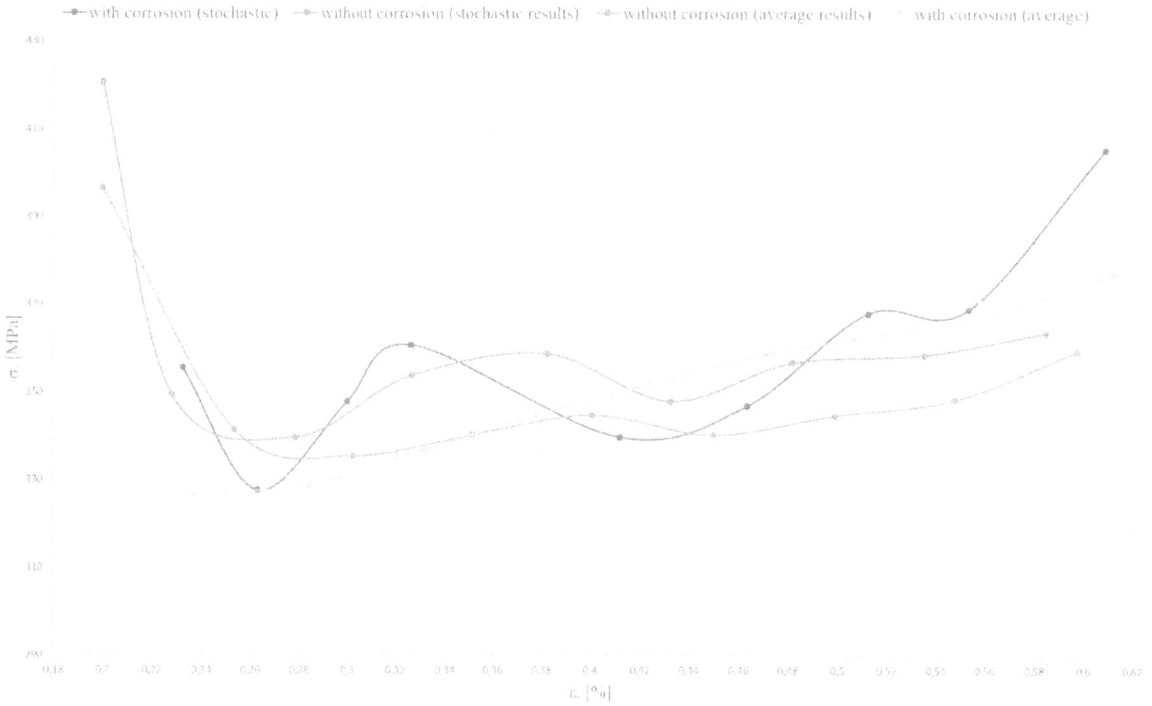

Figure 3.43 Group C yield zone diagrams

Table 3.16 Data on the main points on which the diagram in fig.3.43

type of result	№ points	1	2	3	4	5	6	7	8
stochastic	ε, [%]	0.263	0.300	0.326	0.411	0.463	0.512	0.553	0.608
	σ, [MPa]	327.6	347.8	360.8	339.8	346.9	367.9	368.9	405.3
average	ε, [%]	0.284	0.331	0.377	0.424	0.471	0.518	0.565	0.612
	σ, [MPa]	328.7	336.6	344.7	352.1	358.7	363.1	368.3	376.9

2.4.3. Strain hardening zone

In the studied segment shown in fig.3.43a and table 3.17, it is found that the strip relative deformation has changed significantly, and the stresses are up to 5%. The nonlinear development of mechanical properties is preserved, expressed in a smooth increase in stress and the presence of deformation. It is known that steel has a polycrystalline structure, which means that the volume in the steel contains a large number of small crystals and they have a random orientation. The atoms of the metal are arranged in the crystals in a certain way, building a regular spatial lattice. In the lattice there are forces of interaction between the atoms, a system is formed that is completely defined for each steel. A clearly expressed starting change in the crystal lattice of the corroded steel.

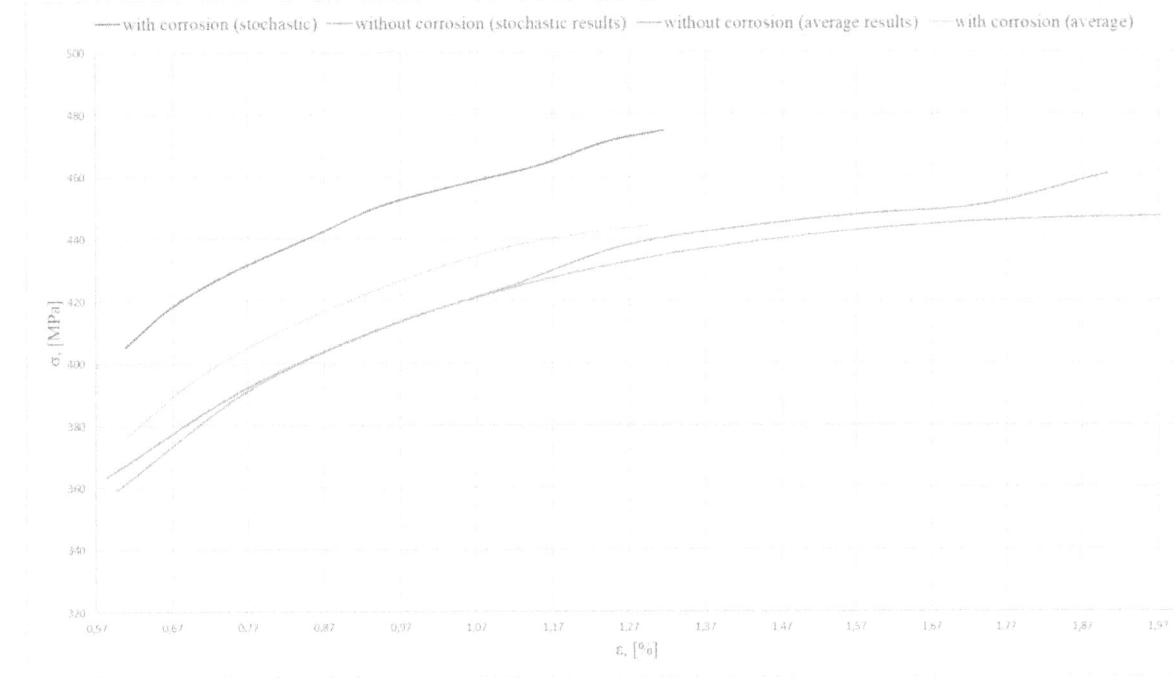

Figure 3.43a Group C strain hardening zone diagrams

Table 3.17 Data on the main points on which the diagram in fig.3.43a

type of result	№ points	1	2	3	4	5	6	7	8
stochastic	ε, [%]	0.667	0.745	0.854	0.945	1.054	1.152	1.241	1.315
	σ, [MPa]	417.9	428.8	440.8	450.8	457.8	463.7	471.3	474.8
average	ε, [%]	0.697	0.783	0.868	0.954	1.039	1.125	1.211	1.296
	σ, [MPa]	394.3	406.8	416.4	425.1	432.3	438.1	441.8	444.2

2.4.4. Necking zone

It is established from fig.3.44 and table 3.18 that the deformation limit decreases to 40%, the failure stress has a negligible change of 5%. The formation of the neck is observed, i.e. relatively large deformations are localized in a small part of the material and a significant reduction in the cross-sectional area occurs. It is established that there is heterogeneity of the corroded steel. The material begins to deform, all places undergo approximately the same amount of deformation and the greater the local stress, the greater the local reduction in the cross-sectional area. It leads to an even greater concentration of stresses, and instability, which causes the formation of the neck. The failure occurs at deformation values characteristic before reaching the tensile strength of uncorroded steel.

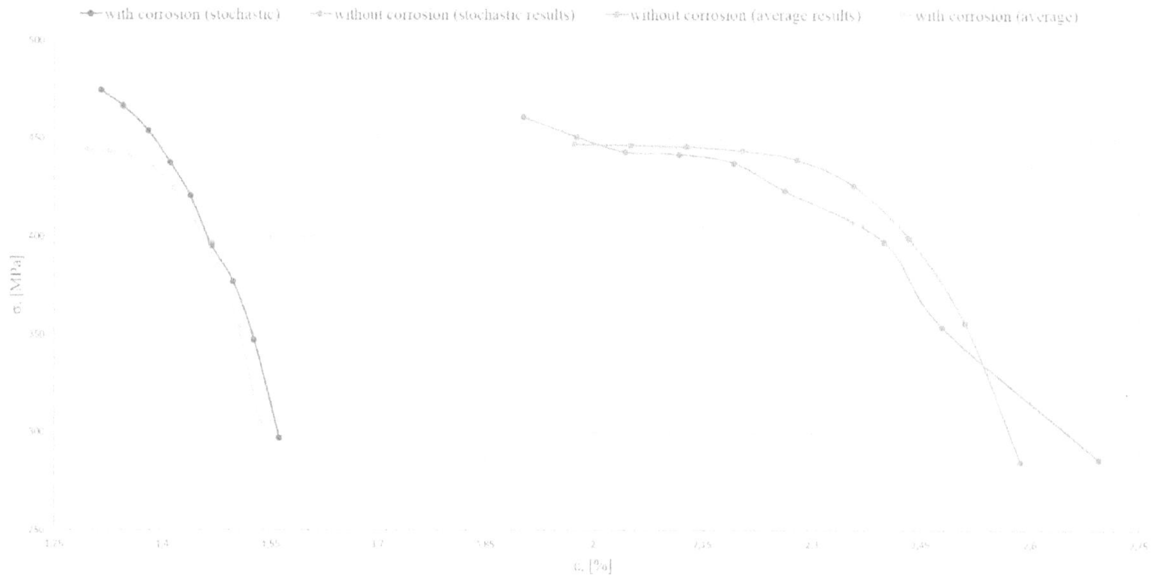

Figure 3.44 Group C necking zone diagrams

Table 3.18 Data on the main points on which the diagram in fig.3.44

type of result	№ points	1	2	3	4	5	6	7	8
stochastic	ε, [%]	1.346	1.381	1.411	1.439	1.468	1.498	1.527	1.562
	σ, [MPa]	466.9	453.9	437.9	420.9	395.4	376.9	347.4	297.4
average	ε, [%]	1.326	1.356	1.386	1.416	1.445	1.475	1.505	1.535
	σ, [MPa]	443.5	441.2	435.5	424.7	408.3	385.8	354.4	304.9

2.4.5. Stress-strain diagram

From the studied working diagram in fig. 3.45 of the corroded material, it is clearly seen that corrosion indicates its influence on the mechanical properties of the studied steel, expressed in the following basic facts:

- reduction of steel ductility by about 40%

- there is an incipient change in the structure of the material, but it continues to remain a ductility material
- pronounced yield strength
- slight increase in strength values up to 5% – 15%
- signs of a significant change in the internal crystal structure of the steel begina
- the modulus of elasticity – decreases by about 30% and begins to not follow Hooke's law, i.e. nonlinearity appears.

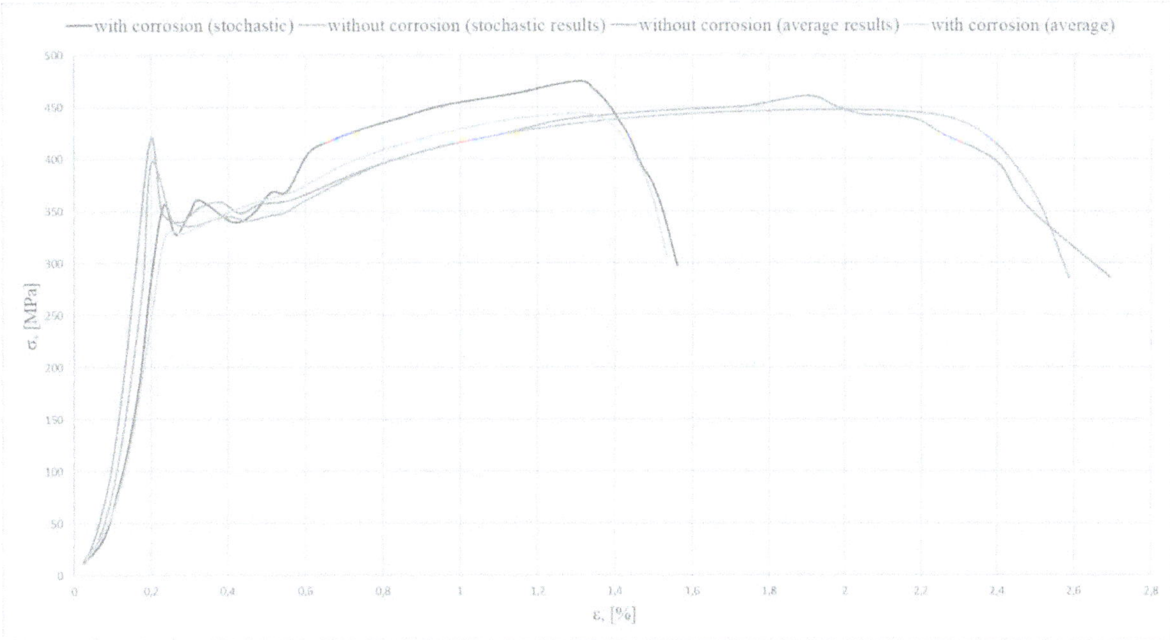

Figure 3.45 Group C stress-strain diagrams

2.5. Study of Group D

Group D consists of 19 corroded steel test specimens with a diameter of 6.00 to 7.00 mm. From fig. 3.46 it is established that the change in the diameter change due to accelerated corrosion is nonlinear as shown in fig.3.46 and does not obey known dependencies. The corrosion effect corresponds as follows: for corrosion category C1 - 34892 months; C2 - 1814 months; C3 - 907 months; C4 - 567 months; C5 - 227 months and CX - 65 months.

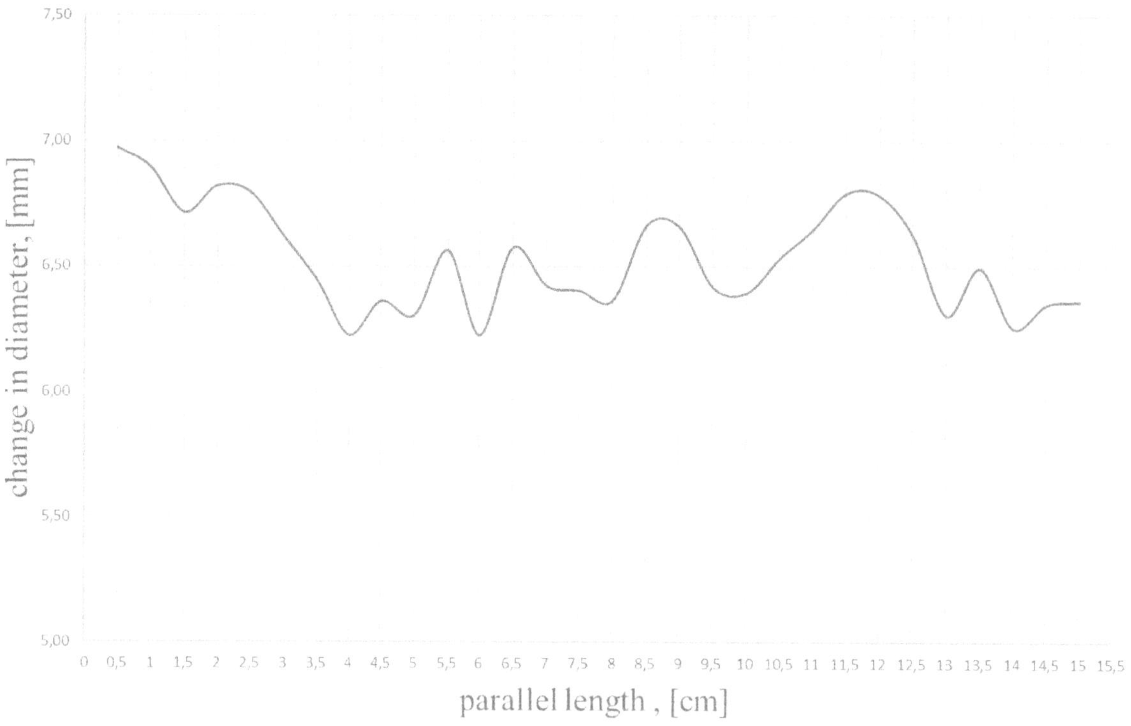

Figure 3.46 Diameter graph after accelerated corrosion in the parallel length – Group D

2.5.1. Elastic zone

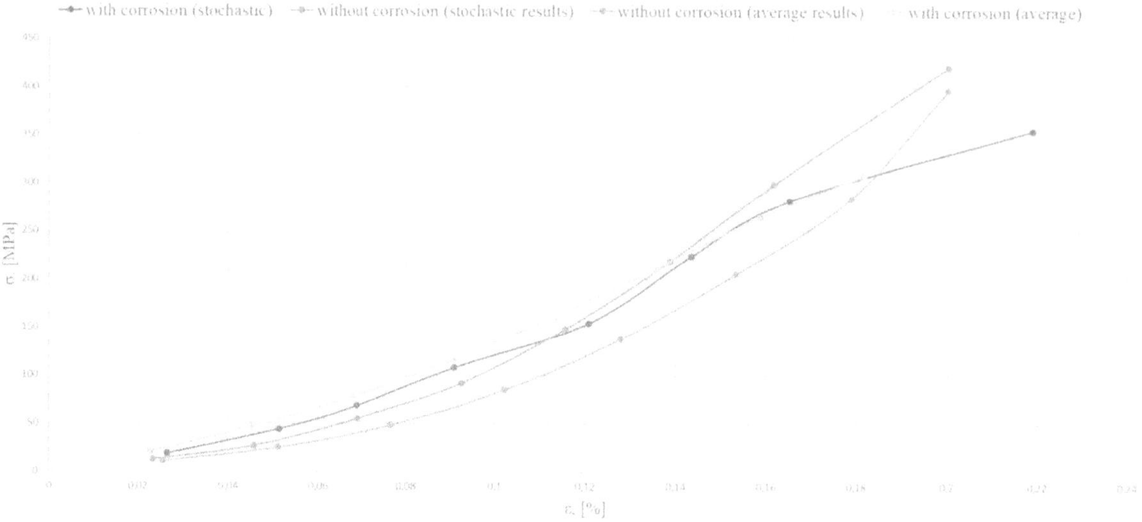

Figure 3.47 Group D elasticity zone diagrams

The studied zone in fig. 3.47 and table 3.19 is characterized by the fact that for steel without corrosion, the section is straight and follows Hooke's law. For corroded elements, it is found that the diagram begins to change, increasing the elastic deformation to 10% and leading to a slight decrease in the stresses in the size to 15%. In general, the zone becomes more elastic and less strong. The boundaries in which the strip is located, the relative deformation does not exceed 0.22%. There is evidence of nonlinear behavior of the modulus of elasticity, although overall it is with a slight decrease.

Table 3.19 Data on the main points on which the diagram in fig.3.47

type of result	№ points	1	2	3	4	5	6	7	8
stochastic	ε, [%]	0.026	0.051	0.069	0.091	0.121	0.143	0.165	0.219
	σ, [MPa]	18.7	44.4	68.5	107.9	154.0	224.0	281.7	355.0
average	ε, [%]	0.023	0.045	0.068	0.091	0.113	0.136	0.159	0.181
	σ, [MPa]	20.9	46.4	77.7	116.3	161.2	212.4	265.0	308.2

2.5.2. Yield zone

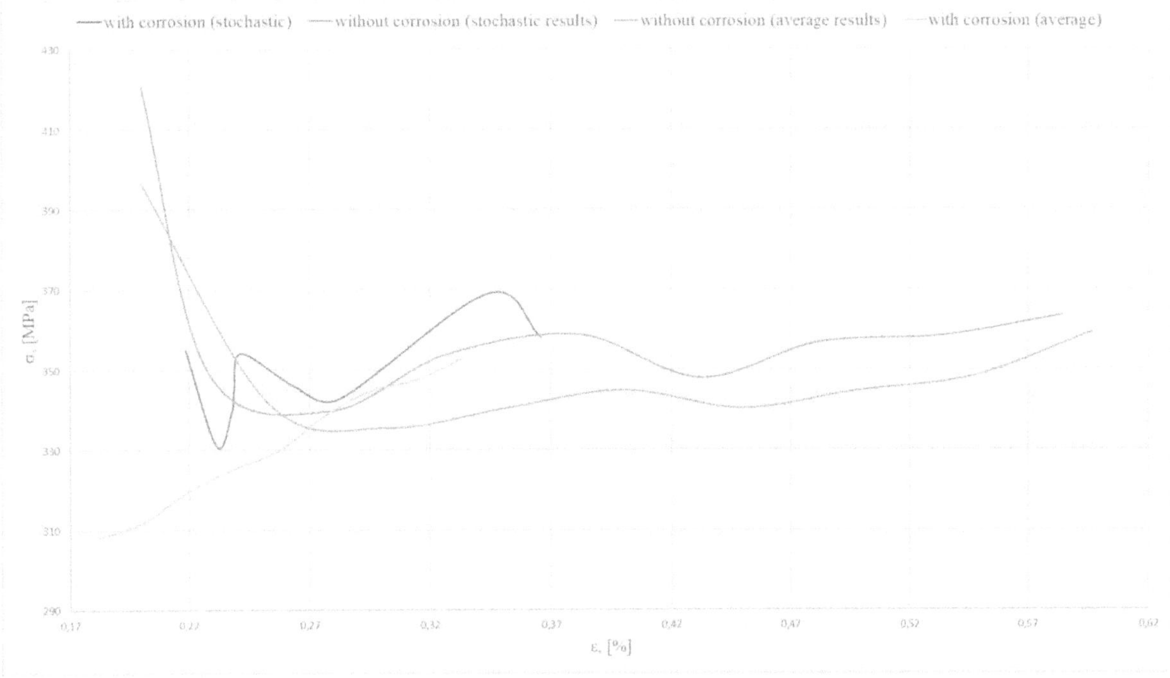

Figure 3.48 Group D yield zone diagrams

In the examined part shown in fig.3.48 and table 3.20, it is established that the diagram is distorted, having a clear expression in the vertical plane, but with the appearance of a length effect in its horizontal part. It is established that the relative deformation has changed, i.e. the yield point exists, but is not categorically definable, and the stresses have decreased within the limits of up to 5%. Circumstances determine that the corroded steel can no longer be classified as a tough material and mean that the failure should occur even at very low values of the deformations. In the part there are plastic deformations that do not change compared to the uncorroded steel.

Table 3.20 Data on the main points on which the diagram in fig.3.48

type of result	№ points	1	2	3	4	5	6	7	8
stochastic	ε, [%]	0.231	0.238	0.239	0.242	0.264	0.283	0.344	0.366
	σ, [MPa]	330.9	339.0	344.9	354.1	345.8	342.8	369.0	357.9
average	ε, [%]	0.200	0.219	0.238	0.258	0.277	0.296	0.315	0.334
	σ, [MPa]	311.7	319.5	325.2	329.8	338.7	344.8	347.5	352.9

2.5.3. Strain hardening zone

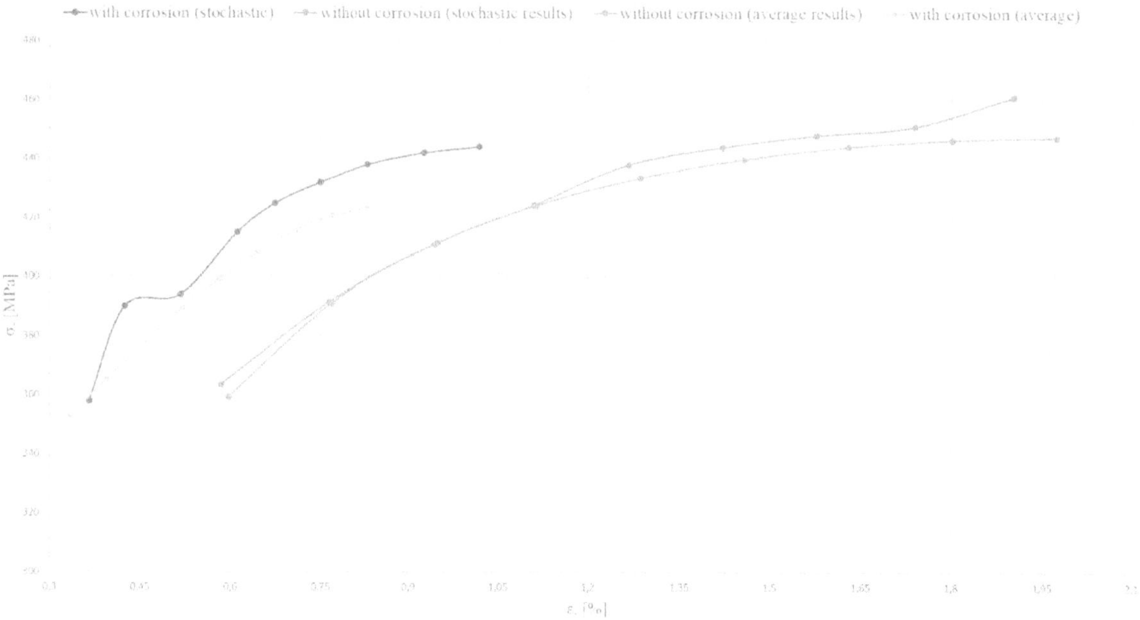

Figure 3.49 Group D strain hardening zone diagrams

In the studied sector in fig.3.49 and table 3.21, it is found that the strip relative deformation has changed significantly, within the limits of 40% to 50%. And the stresses follow a slight decrease to 5%. The nonlinear development of the mechanical properties is preserved, expressed in a sharp increase in the stress relative to the available deformation. It is known that steel has a polycrystalline structure, which means that the volume in the steel contains a large number of small crystals that have a random orientation. The atoms of the metal are arranged in the crystals in a certain way, building a regular spatial lattice. In the lattice there are forces of interaction between them, a system is formed that is completely defined for each steel. A change in the crystal lattice of the corroded steel is clearly expressed.

Table 3.21 Data on the main points on which the diagram in fig.3.49

type of result	№ points	1	2	3	4	5	6	7	8
stochastic	ε, [%]	0.425	0.517	0.611	0.674	0.749	0.831	0.925	1.018
	σ, [MPa]	390.0	394.0	415.0	425.0	432.0	438.0	442.0	444.0
average	ε, [%]	0.396	0.458	0.520	0.583	0.645	0.707	0.769	0.832
	σ, [MPa]	365.5	377.8	389.3	399.6	408.2	415.8	420.7	423.7

2.5.4. Necking zone

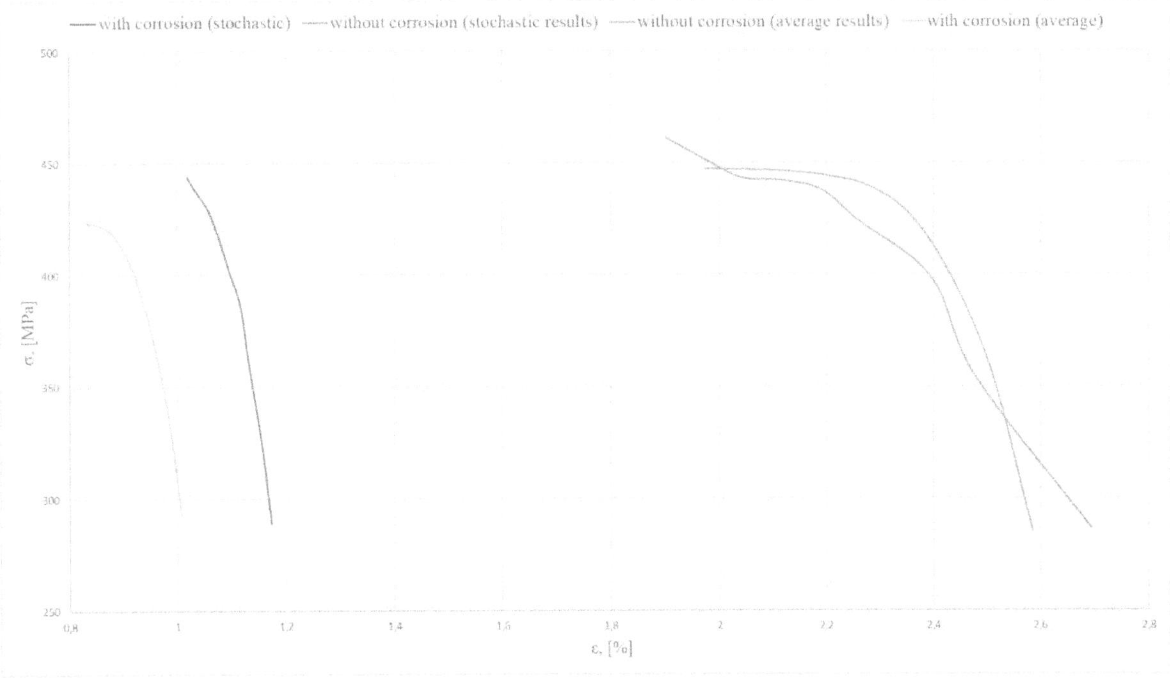

Figure 3.50 Group D necking zone diagrams

From fig. 3.50 and table 3.22, it is determined that the deformation limit decreases to 55%, the failure stress has a negligible change of 5%. The formation of the neck is observed, i.e. relatively large deformations are localized in a small part of the material and a significant reduction in the cross-sectional area occurs. It is found that there is heterogeneity of the corroded steel. As this material begins to deform, all places undergo approximately the same amount of deformation and the greater the local stress, the greater the local reduction in the cross-sectional area, which in turn leads to an even greater concentration of stresses, leading to instability, which causes the formation of the neck. The failure occurs at deformation values characteristic shortly after the tensile strength of corroded steel is reached.

Table 3.22 Data on the main points on which the diagram in fig.3.50

type of result	№ points	1	2	3	4	5	6	7	8
stochastic	ε, [%]	1.033	1.057	1.077	1.097	1.117	1.133	1.156	1.173
	σ, [MPa]	438.0	429.0	416.0	401.0	386.0	357.9	323.9	288.9
average	ε, [%]	0.854	0.876	0.898	0.920	0.942	0.964	0.986	1.008
	σ, [MPa]	422.6	419.3	412.0	399.9	382.9	360.8	332.8	292.7

2.5.5. Stress-strain diagram

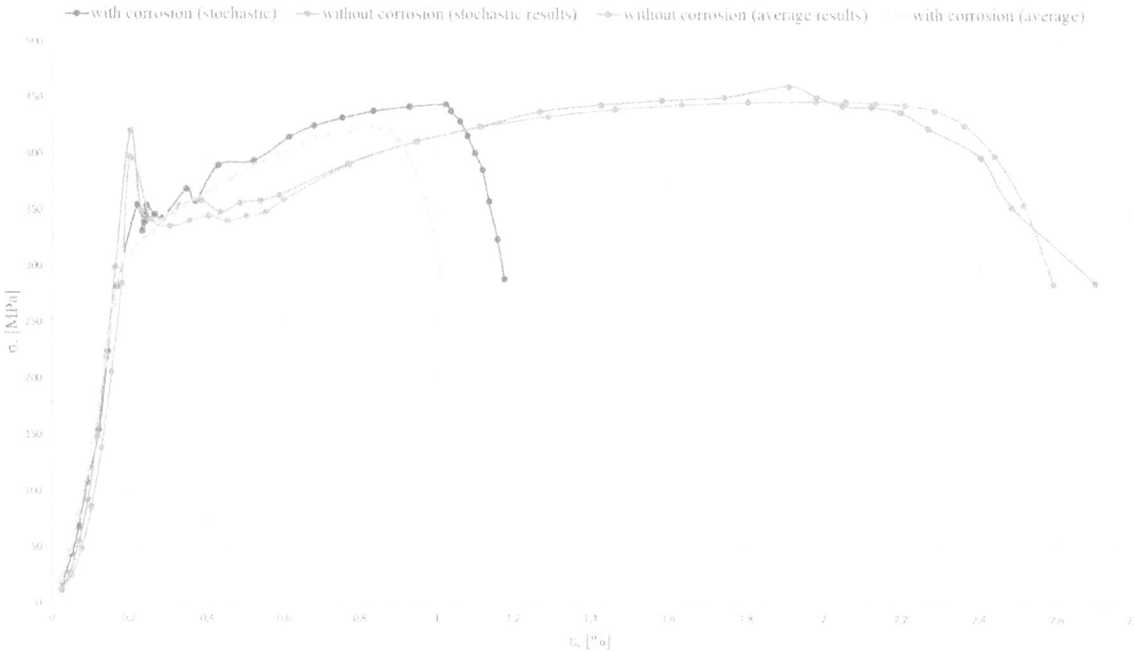

Figure 3.51 Group D stress-strain diagrams

From the examined working diagram in fig. 3.51 of the corroded material, it is seen that corrosion indicates its influence on the mechanical properties of the studied steel, expressed in the following basic facts:

- reduction of steel ductility by about 55%
- there is a change in the structure of the material and it cannot be classified as a ductility material.
- weakly expressed yield strength

- slight increase in strength values up to 5% – 15%
- changes the internal crystal structure of the steel
- the modulus of elasticity – decreases by about 35%, has a nonlinear nature and does not obey Hooke's law

2.6. Study of Group E

Group E consists of 19 corroded steel test specimens with a diameter ranging from 5.00 to 6.00 mm after application of accelerated corrosion.

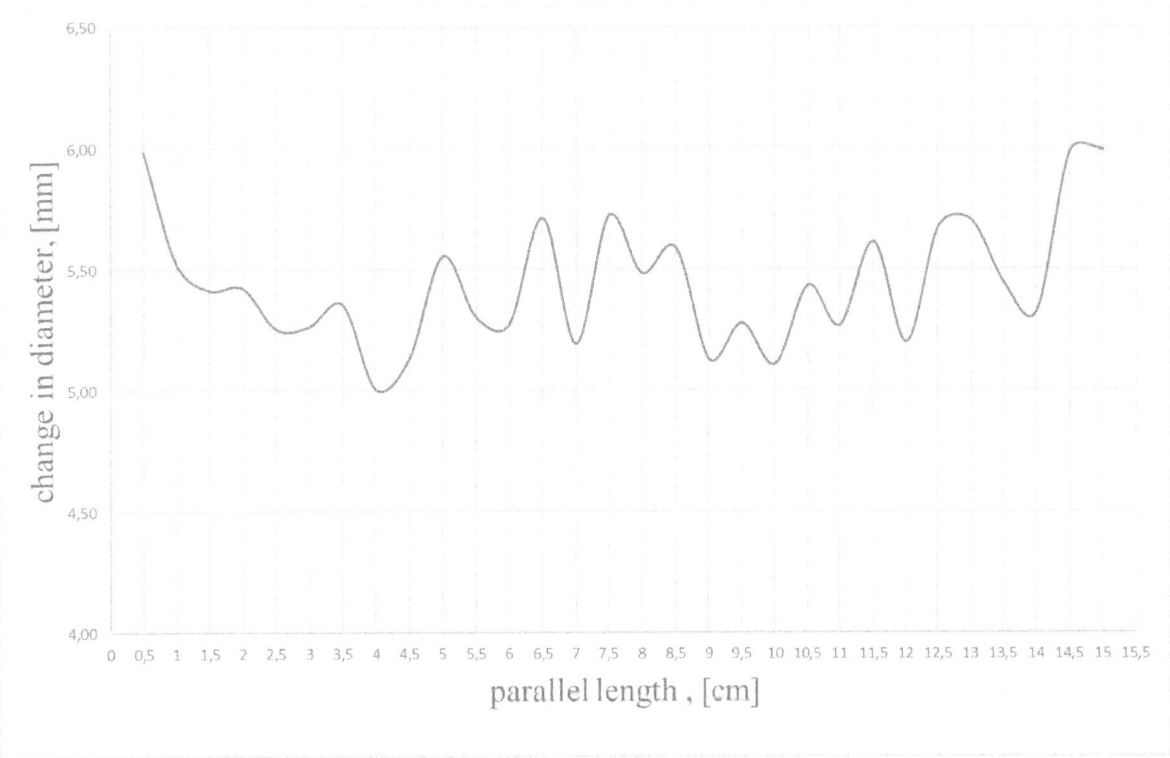

Figure 3.52 Diameter graph after accelerated corrosion in the parallel length – Group E

On fig.3.52 shows the change in the diameter change due to accelerated corrosion is nonlinear. It does not obey any known dependencies, which fact brings us as close as possible to the real conditions of impact indicating corrosion on the cross section in a corroded steel element. The application of the ISO 9223 standard gives reason to establish that group D corresponds to corrosion impact: for corrosion category C1 - 46154 months; C2 - 2400 months; C3 - 1200 months; C4 - 750 months; C5 - 300 months and CX - 86 months.

2.6.1. Elastic zone

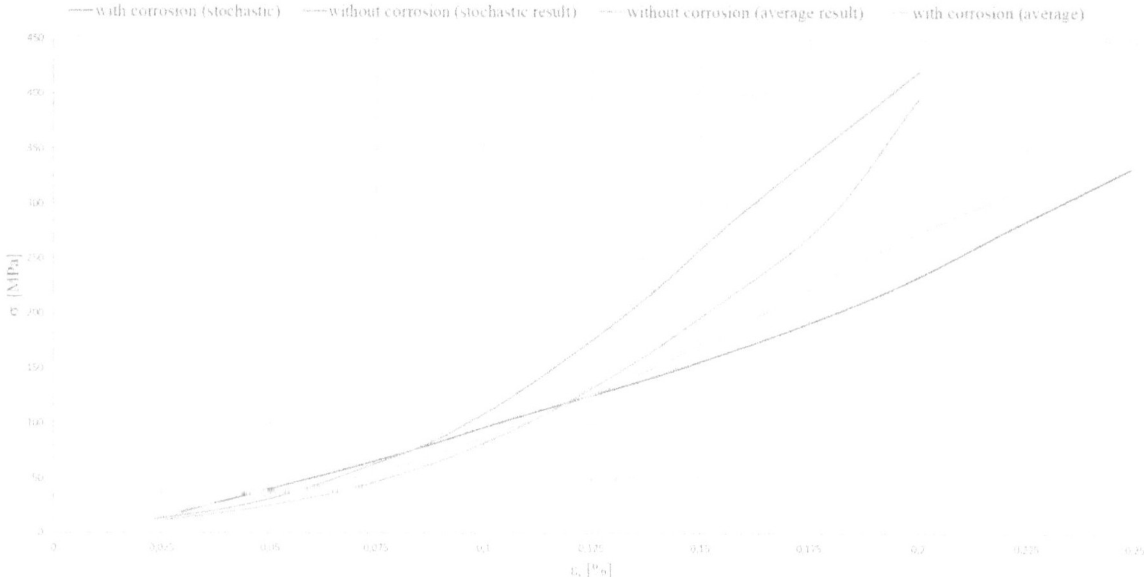

Figure 3.53 Group E elasticity zone diagrams

Table 3.23 Data on the main points on which the diagram in fig.3.53

type of result	№ points	1	2	3	4	5	6	7	8
stochastic	ε, [%]	0.029	0.051	0.077	0.106	0.149	0.191	0.223	0.249
	σ, [MPa]	18.7	41.2	68.9	104.0	154.7	218.0	282.4	332.6
average	ε, [%]	0.028	0.056	0.084	0.112	0.140	0.167	0.195	0.223
	σ, [MPa]	19.9	38.1	65.0	104.0	151.4	206.4	265.1	313.4

The studied zone in Fig.3.53 and Table 3.23 is characterized by the fact that for steel without corrosion, the section is straight and follows Hooke's law. For corroded elements, it is found that the diagram begins to change, increasing the elastic deformation to 20% and leading to a decrease in the stresses in the size to 20%. In general, the zone becomes more elastic and less strong. The boundaries in which the strip is the relative deformation do not exceed 0.25%. Non-linear behavior of the elastic modulus appears, it decreases to 35%.

2.6.2. Yield zone

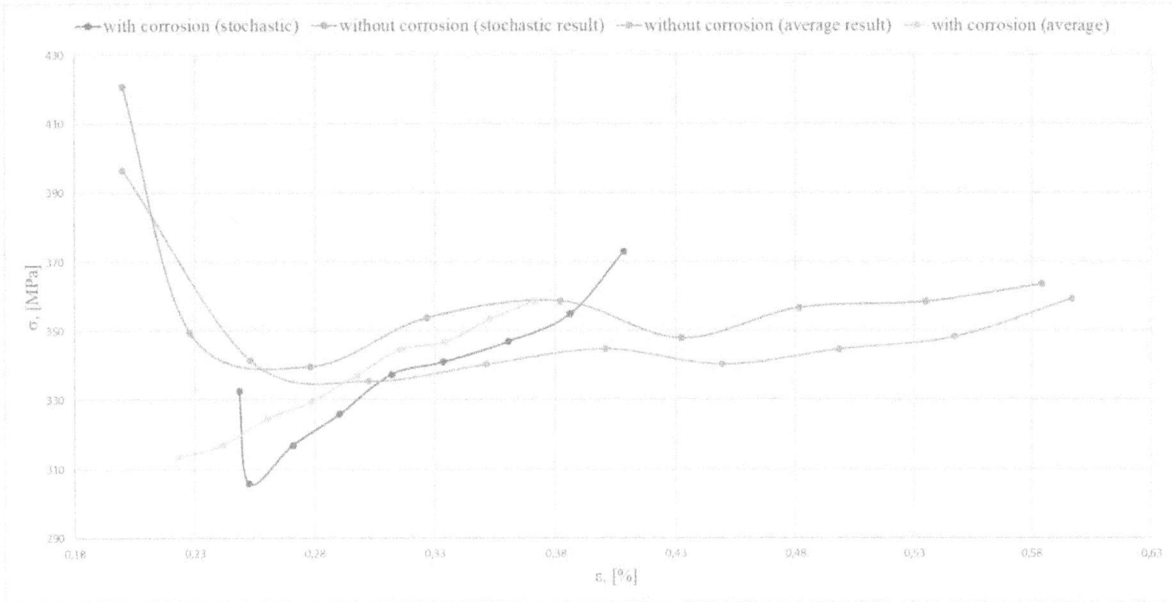

Figure 3.54 Group E yield zone diagrams

Table 3.24 Data on the main points on which the diagram in fig.3.54

type of result	№ points	1	2	3	4	5	6	7	8
stochastic	ε, [%]	0.253	0.271	0.290	0.312	0.334	0.361	0.386	0.408
	σ, [MPa]	305.9	316.9	325.9	337.4	341.0	346.9	354.9	372.9
average	ε, [%]	0.242	0.260	0.279	0.297	0.316	0.334	0.353	0.371
	σ, [MPa]	317.0	324.7	329.5	337.2	344.7	346.9	353.3	358.6

In the investigated area of fFig.3.54 and table 3.24, it is established that the diagram is distorted, having a clear expression in the vertical plane, but with the appearance of a length effect in its horizontal part. It is established that the relative deformation has changed, i.e. the yield point exists, but is not categorically definable, the stresses have a slight decrease in the range of up to 10%. Circumstances determine that the corroded steel can no longer be classified as a tough material and mean that the failure should occur even at low values of deformations. There are plastic deformations that do not change compared to the uncorroded steel.

2.6.3. Strain hardening zone

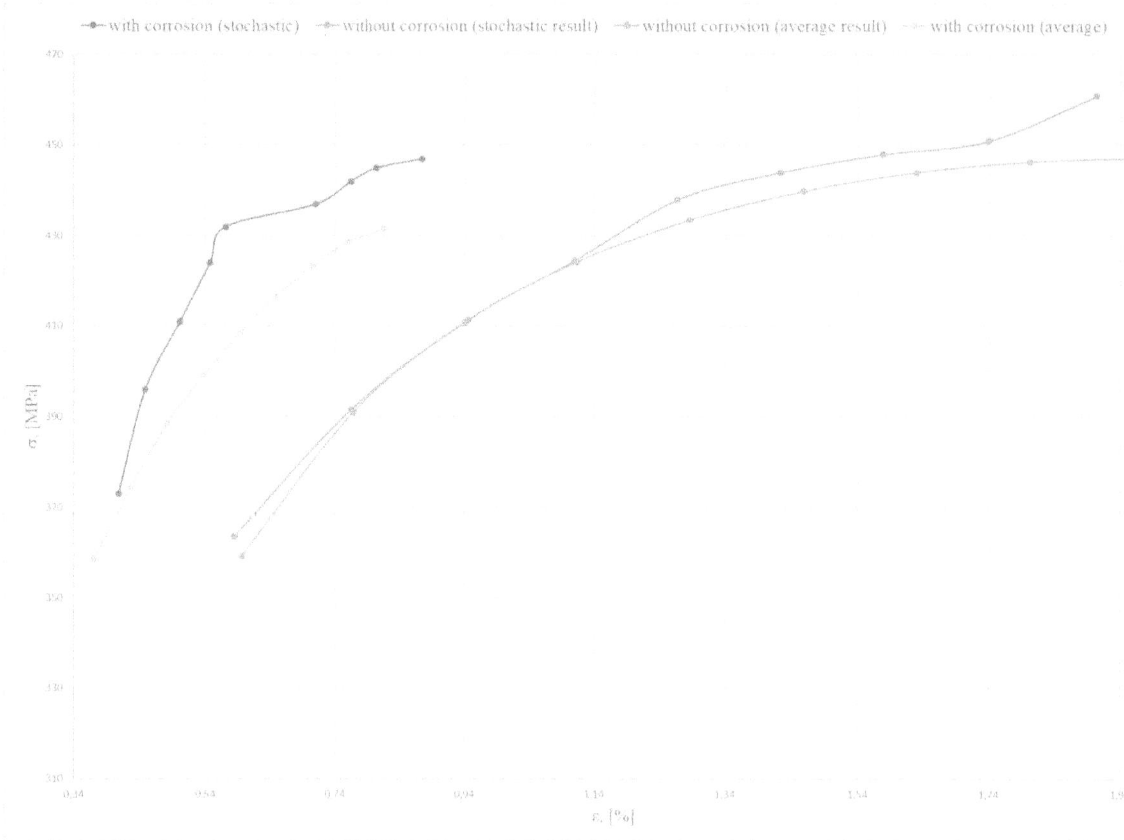

Figure 3.55 Group E strain hardening zone diagrams

Table 3.25 Data on the main points on which the diagram in fig.3.55

type of result	№ points	1	2	3	4	5	6	7	8
stochastic	ε, [%]	0.449	0.502	0.547	0.571	0.710	0.764	0.805	0.875
	σ, [MPa]	395.9	410.9	423.9	431.9	436.9	441.9	444.9	446.9
average	ε, [%]	0.427	0.483	0.538	0.594	0.650	0.705	0.761	0.817
	σ, [MPa]	374.6	388.5	399.4	408.5	416.6	423.2	428.6	431.4

From fig.3.55 and table 3.25, it is found that the strip relative deformation has changed, within the limits of 50% to 60%. The stresses follow a decrease of up to 10%. The nonlinear development of the mechanical properties is preserved, expressed in a sharp increase in the stress relative to the presence of deformation. It is known that steel has a polycrystalline structure, which means that the volume in the steel contains a large number of small crystals that have a random orientation. The atoms of the metal are arranged in the crystals in a certain way, building a regular spatial lattice. In the lattice there are forces of interaction between the atoms, they form a system that is completely defined for each steel. A clearly expressed change in the crystal lattice of the corroded steel.

2.6.4. Necking zone

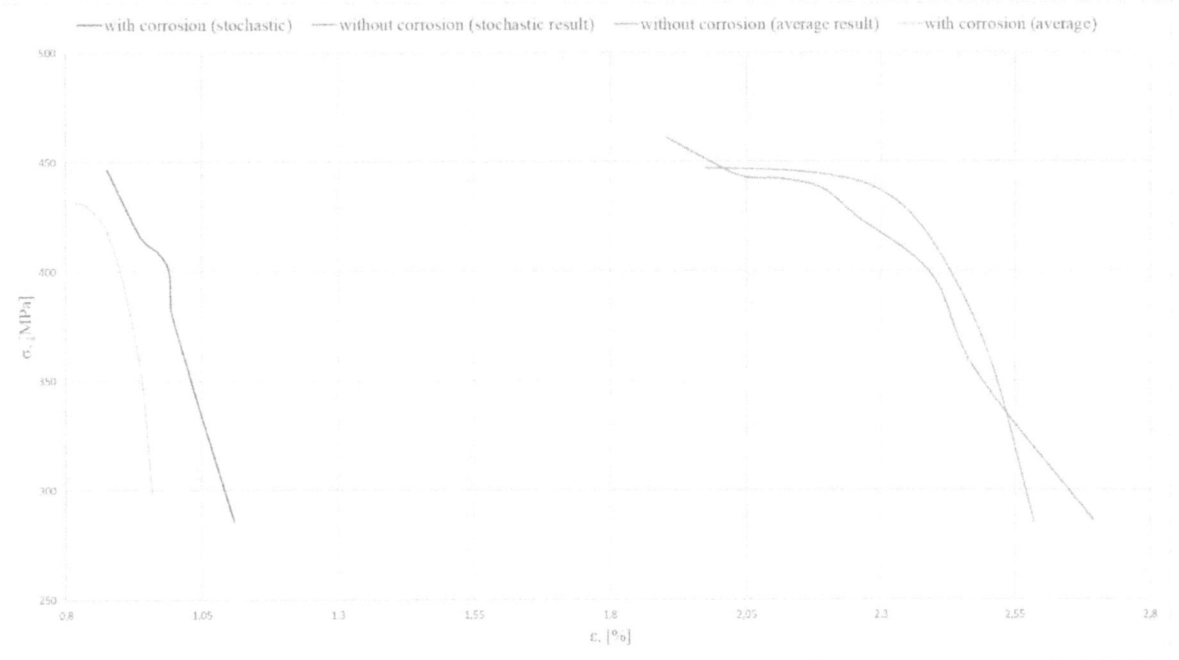

Figure 3.56 Group E necking zone diagrams

Table 3.26 Data on the main points on which the diagram in fig.3.56

type of result	№ points	1	2	3	4	5	6	7	8
stochastic	ε, [%]	0.936	0.968	0.988	0.992	0.994	1.019	1.049	1.109
	σ, [MPa]	415.9	408.9	400.9	387.9	380.9	359.9	334.9	285.9
average	ε, [%]	0.834	0.852	0.870	0.887	0.905	0.923	0.941	0.958
	σ, [MPa]	430.5	426.2	420.9	409.9	394.3	374.0	347.7	298.3

In the studied sector of fig. 3.56 and table 3.26, it is determined that the deformation limit decreases to 60%, the failure stress has a negligible change of 5%. The formation of the neck is observed, i.e. relatively large deformations are localized in a small part of the material and a significant reduction in the cross-sectional area occurs. It is found that there is heterogeneity of the corroded steel. As the material begins to deform, all places undergo approximately the same amount of deformation. The greater the local stress, the greater the local reduction in the cross-sectional area. It leads to an even greater concentration of stresses, which leads to instability, which causes the formation of the neck. The failure occurs at deformation values characteristic of the initial part of the diagram in the hardening zone before reaching the tensile strength of uncorroded steel.

2.6.5. Stress-strain diagram

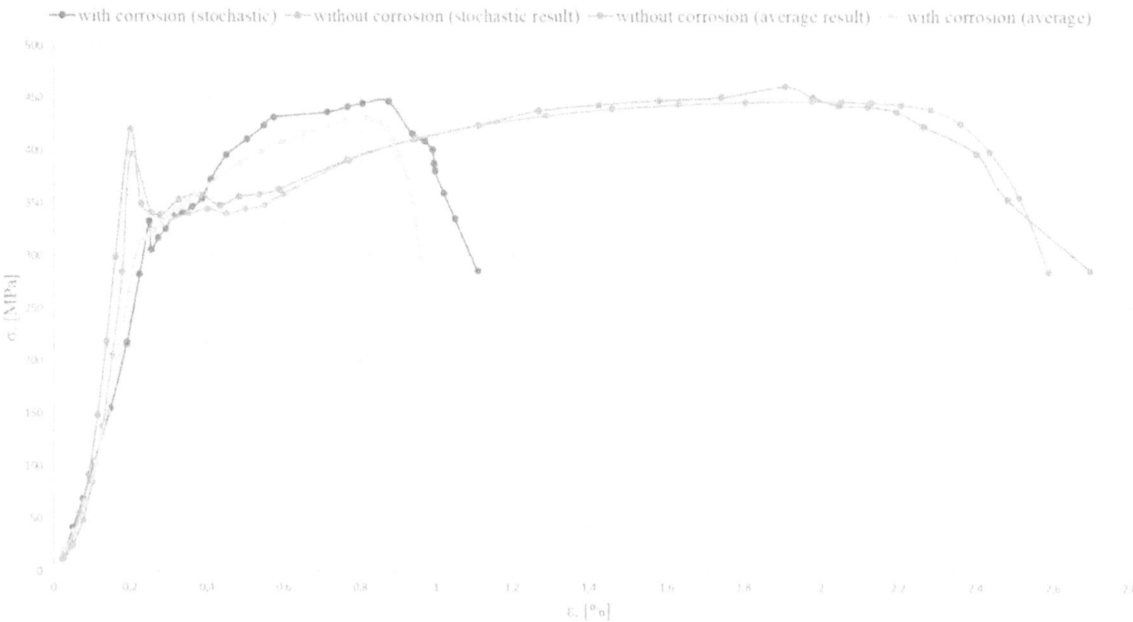

Figure 3.57 Group E stress-strain diagrams

From the studied stress-strain diagram in fig. 3.57 of the corroded material, it is clearly established that corrosion indicates its influence on the mechanical properties of the studied steel, expressed in the following basic facts:

- reduction of steel ductility by about 60%
- there is a change in the structure of the material and it cannot be classified as a ductility material
- weakly expressed yield strength
- slight increase in strength values up to 5% – 15%
- change in the internal crystal structure of the steel
- the modulus of elasticity – decreases by about 35%, has a nonlinear nature and does not obey Hooke's law.

2.7. Study of Group F

Group F consists of 17 corroded steel test specimens with a diameter ranging from 4.00 to 5.00 mm after application of accelerated corrosion.

Figure 3.58 Diameter graph after accelerated corrosion in the parallel length – Group F

It is established from fig.3.58 that the change in the diameter change due to accelerated corrosion is nonlinear and does not obey known dependencies, which fact brings us as close as possible to the real conditions of impact. This affects the cross-section in a corroded steel element. The application of the ISO 9223 standard gives grounds to establish that group F corresponds to corrosion impact as follows: for corrosion category C1 - 54000 months; C2 - 2808 months; C3 - 1404 months; C4 - 878 months; C5 - 351 months and CX - 100 months.

2.7.1. Elastic zone

In fig.3.59 and table 3.27 are the data of a section and it is characterized that for steel without corrosion, the section is straight and follows Hooke's law. In corroded elements it is established that the diagram begins to change, increasing the elastic deformation to 10% and leading to a slight decrease in the stresses in the size to 5%. In general, the zone becomes more elastic and less strong. The boundaries in which the strip is the relative deformation do not exceed 0.21%. It can be clearly established nonlinear behavior of the modulus of elasticity, although in general a decrease is noted.

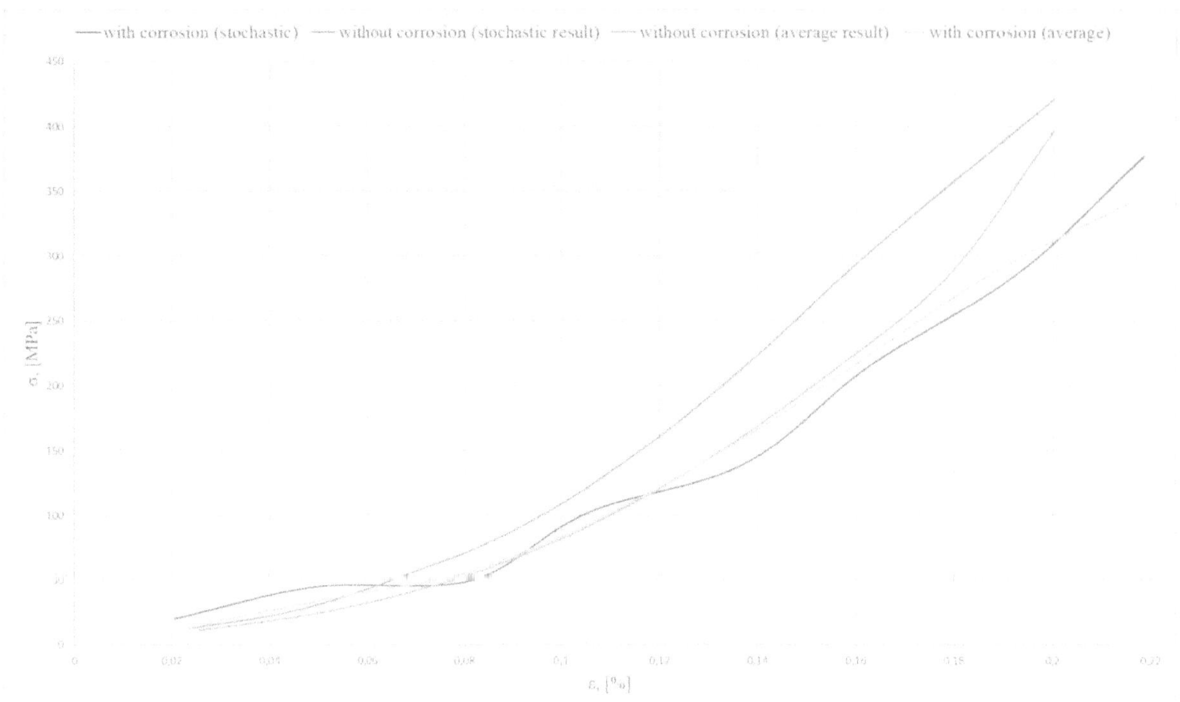

Figure 3.59 Group F elasticity zone diagrams

Table 3.27 Data on the main points on which the diagram in fig.3.59

type of result	№ points	1	2	3	4	5	6	7	8
stochastic	ε, [%]	0.021	0.049	0.082	0.105	0.137	0.162	0.193	0.218
	σ, [MPa]	20.0	44.7	49.8	101.9	139.4	213.4	286.8	376.9
average	ε, [%]	0.027	0.054	0.081	0.108	0.135	0.162	0.188	0.215
	σ, [MPa]	16.4	36.5	55.8	97.0	153.4	220.5	288.5	340.3

2.7.2. Yield zone

In the studied segment of fig. 3.60 and table 3.28, it is established that the diagram is distorted, having a clear expression in the vertical plane, but with the appearance of a length effect in its horizontal part. It is established that the relative deformation has changed, i.e. the yield point exists, but is not categorically definable, but the stresses have increased in the range of up to 10%. Circumstances determine that the corroded steel can no longer be classified as a tough material and mean that the failure should occur even at low values of the deformations. In the part there are plastic deformations that do not change compared to the uncorroded steel.

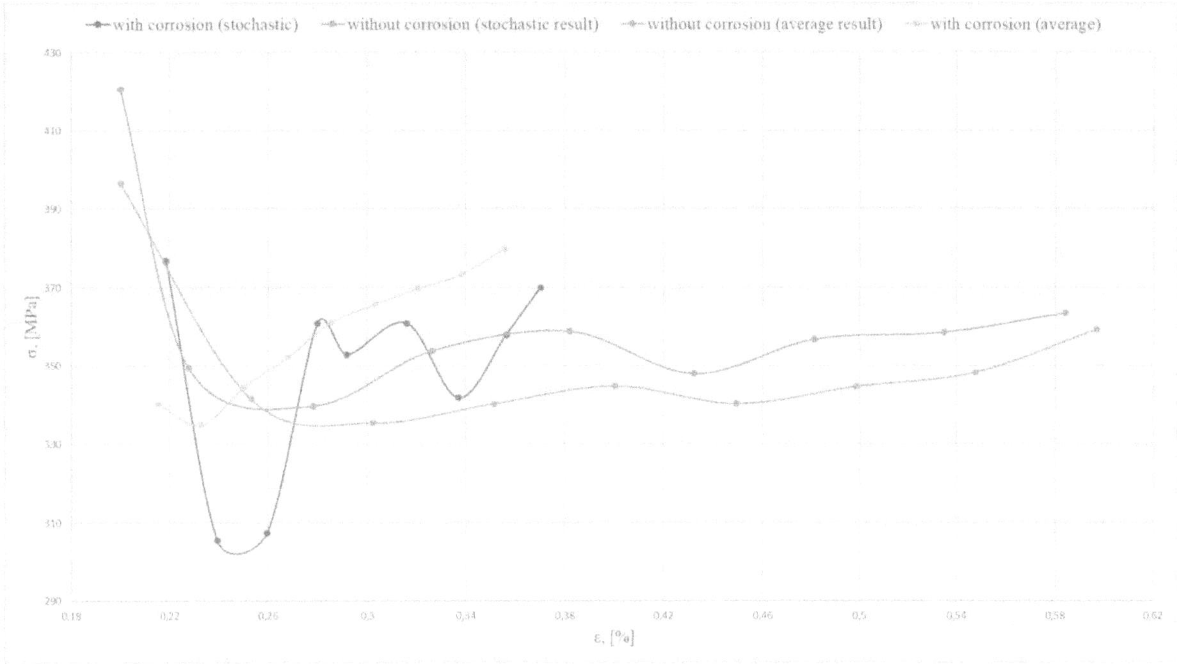

Figure 3.60 Group F yield zone diagrams

Table 3.28 Data on the main points on which the diagram in fig.3.60

type of result	№ points	1	2	3	4	5	6	7	8
stochastic	ε, [%]	0.240	0.259	0.280	0.292	0.316	0.337	0.356	0.370
	σ, [MPa]	305.5	307.4	360.8	352.8	360.8	341.8	357.8	369.8
average	ε, [%]	0.233	0.251	0.268	0.286	0.303	0.321	0.338	0.356
	σ, [MPa]	334.9	344.7	352.2	361.1	365.6	369.7	373.3	379.9

2.7.3. Strain hardening zone

In the studied area on fig.3.61 and table 3.29, it is established that the strip relative deformation has changed significantly, within the limits of 50% to 60%. The stresses have decreased by up to 10%. The nonlinear development of the mechanical properties is preserved, expressed in a sharp increase in the stress versus the deformation. It is known that steel has a polycrystalline structure, which means that the volume in the steel contains a large number of small crystals that have a random orientation. The atoms of the metal are located in the crystals in a certain way, building a regular spatial lattice.

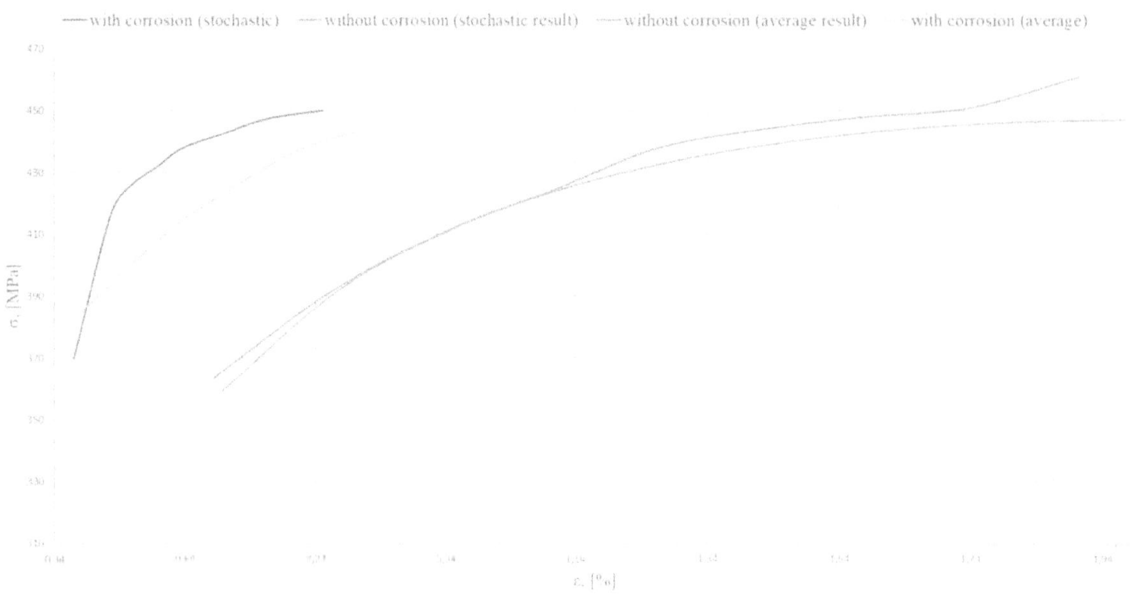

Figure 3.61 Group F strain hardening zone diagrams

Table 3.29 Data on the main points on which the diagram in fig.3.61

type of result	№ points	1	2	3	4	5	6	7	8
stochastic	ε, [%]	0.429	0.459	0.500	0.538	0.606	0.644	0.682	0.750
	σ, [MPa]	416.9	425.9	431.9	437.9	442.9	445.9	447.9	449.9
average	ε, [%]	0.411	0.467	0.522	0.578	0.634	0.689	0.745	0.800
	σ, [MPa]	390.7	401.9	412.3	420.3	427.7	434.6	439.9	443.0

2.7.4. Necking zone

In the studied sector in fig. 3.62 and table 3.30, it is determined that the deformation limit decreases to 60%, the failure stress has a negligible change of 5%. The formation of a neck is observed, i.e. relatively large deformations are localized in a small part of the material and a significant reduction in the cross-sectional area occurs. It is found that there is heterogeneity of the corroded steel. The material begins to deform, all places undergo approximately the same amount of deformation. The greater the local stress, the greater the local reduction in the cross-sectional area. It leads to even greater stress concentration, instability and the formation of a neck. The failure occurs at deformation values characteristic of the initial part of the diagram in the strengthening zone before reaching the tensile strength of non-corroded steel.

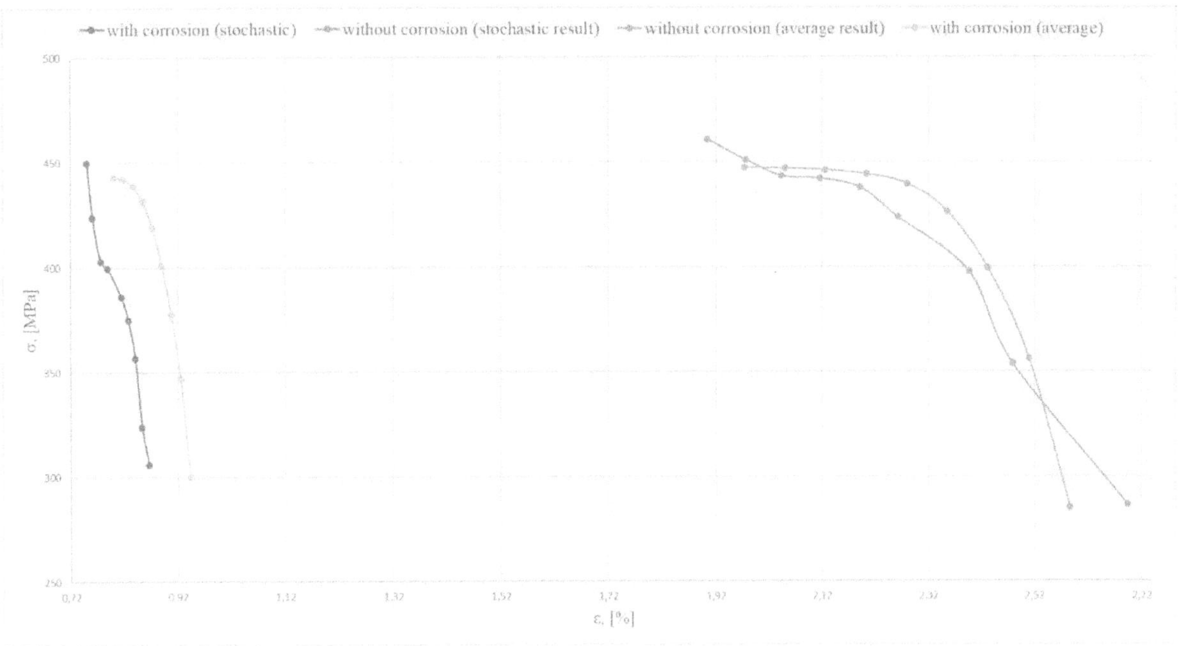

Figure 3.62 Group F necking zone diagrams

Table 3.30 Data on the main points on which the diagram in fig.3.62

type of result	№ points	1	2	3	4	5	6	7	8
stochastic	ε, [%]	0.761	0.776	0.789	0.815	0.827	0.840	0.853	0.866
	σ, [MPa]	423.9	402.8	399.8	385.8	374.8	356.8	323.8	305.9
average	ε, [%]	0.818	0.836	0.854	0.872	0.890	0.908	0.926	0.944
	σ, [MPa]	441.7	438.7	431.5	419.0	401.1	377.4	346.7	300.3

2.7.5. Stress-strain diagram

From the studied stress-strain diagram in fig. 3.63 of the corroded material, it is shown that corrosion indicates its influence on the mechanical properties of the studied steel, expressed in the following basic facts:

- reduction of steel ductility by about 60%
- there is a change in the structure of the material and it cannot be classified as a ductility material
- weakly expressed yield strength
- slight increase in strength values up to 5% – 15%
- changes in the internal crystal structure of the steel

- the modulus of elasticity – decreases by about 25%, has a nonlinear nature and does not obey Hooke's law.

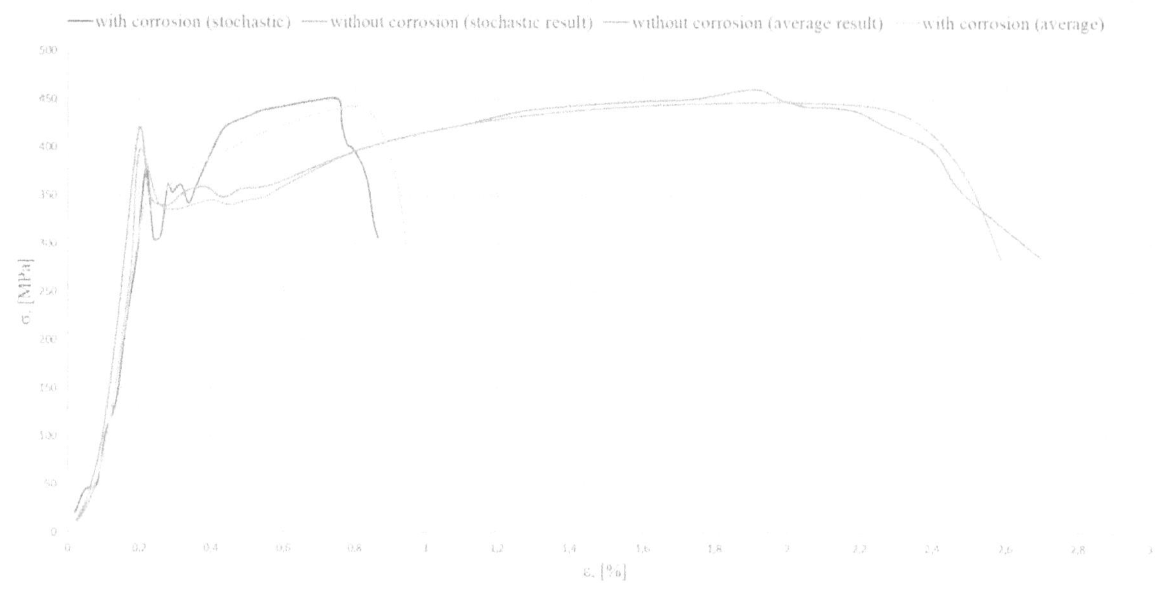

Figure 3.63 Group F stress-strain diagrams

2.8. Study of Group G

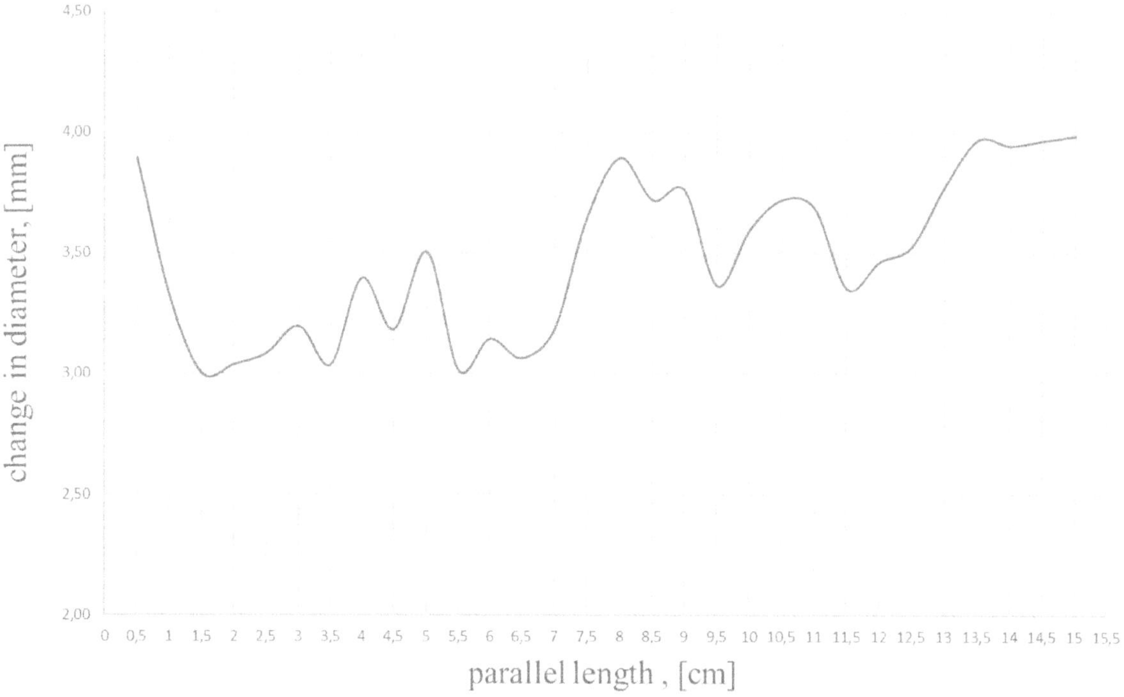

Figure 3.64 Diameter graph after accelerated corrosion in the parallel length – Group G

Group G consists of 9 corroded steel test specimens with a diameter varying after application of accelerated corrosion from 3.00 to 4.00 mm. From fig. 3.64 it is established that the change in the diameter change as a result of accelerated corrosion is nonlinear and does not obey known dependencies. The corrosion effect corresponds to: for corrosion category C1 - 64615 months; C2 - 3360 months; C3 - 1680 months; C4 - 1050 months; C5 - 420 months and CX - 120 months.

2.8.1. Elastic zone

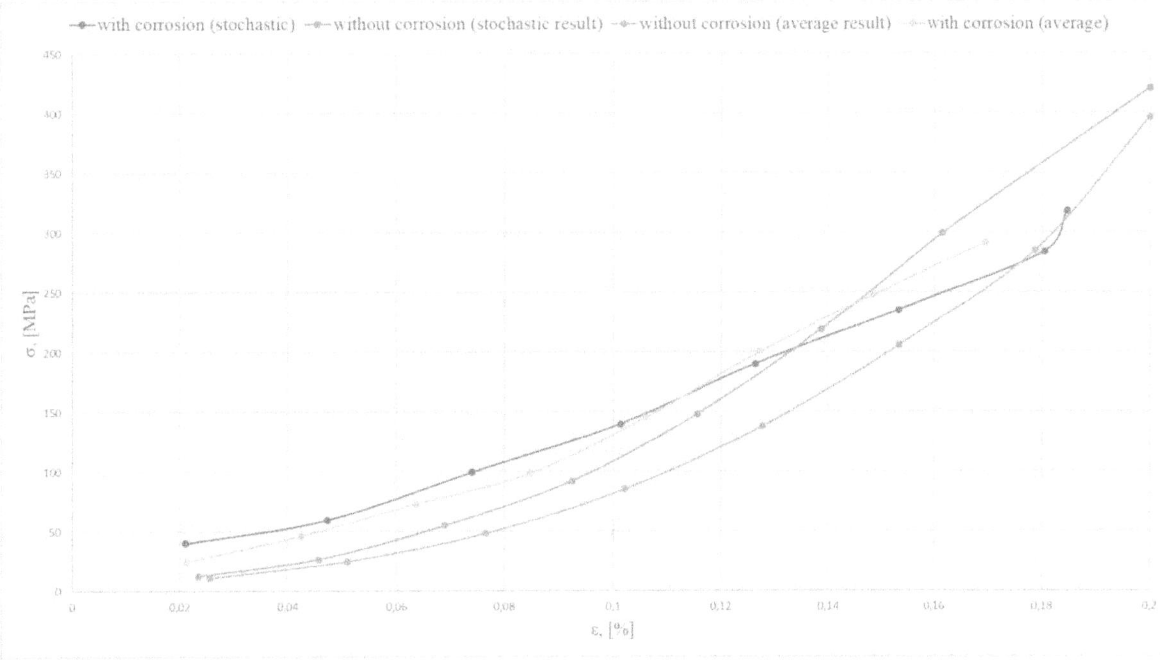

Figure 3.65 Group G elasticity zone diagrams

Table 3.31 Data on the main points on which the diagram in fig.3.65

type of result	№ points	1	2	3	4	5	6	7	8
stochastic	ε, [%]	0.021	0.047	0.074	0.101	0.126	0.153	0.181	0.185
	σ, [MPa]	40.0	59.7	99.4	139.9	189.9	234.9	283.9	317.8
average	ε, [%]	0.021	0.042	0.064	0.085	0.106	0.127	0.148	0.170
	σ, [MPa]	24.4	46.1	72.5	98.4	145.0	200.2	248.0	291.2

The studied zone in fig.3.65 and table 3.31, it is found that for steel without corrosion, the section is straight and follows Hooke's law. For corroded elements, it is found that the diagram begins to change, such as a decrease in elastic deformation to 10% and a significant decrease in stresses in the size to 25%. The zone begins to lose its elasticity and strength properties. The boundaries in which the strip is the relative deformation do not exceed 0.18%. A nonlinear behavior of the modulus of elasticity can be clearly established. A decrease is registered.

2.8.2. Yield zone

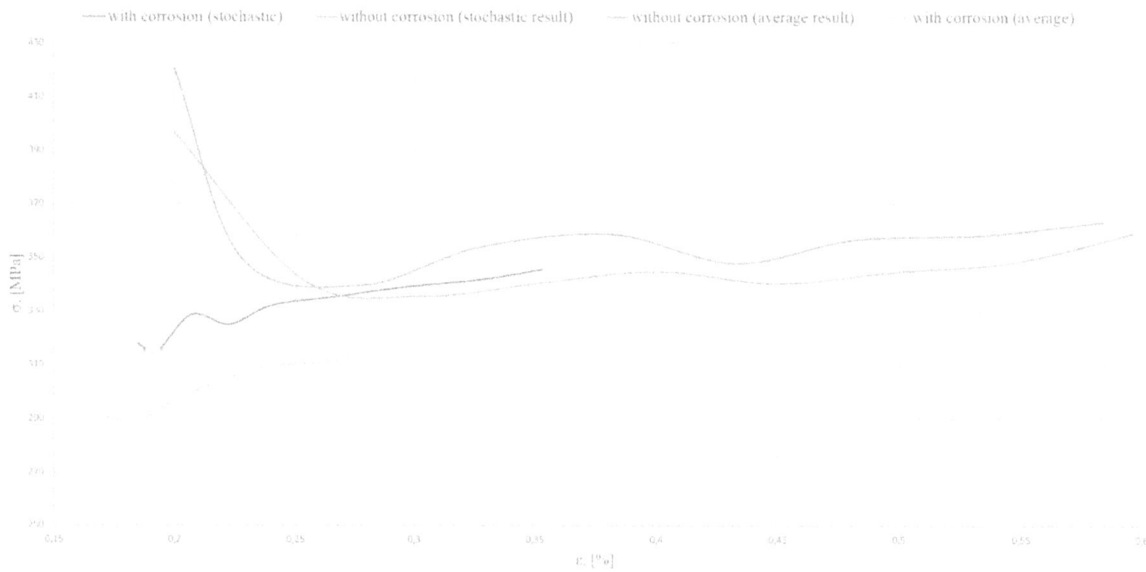

Figure 3.66 Group G yield zone diagrams

In the studied section of fig.3.66 and table 3.32, it is established that the diagram is distorted, having a clear expression in the vertical direction, but with the appearance of a minimal length effect in its horizontal part. It is established that the relative deformation has changed, i.e. the yield point does not exist, and cannot be categorically defined, but the stresses have decreased within the limits of up to 10%. Circumstances determine that the corroded steel can no longer be classified as a tough material. It has passed into another phase, i.e. it begins to acquire the behavior of a brittle material, the destruction should occur even at low values of the tensile deformations. In the part there are plastic deformations that do not change compared to the uncorroded steel.

Table 3.32 Данни на основните points по който е направена диаграма на 3.66

type of result	№ points	1	2	3	4	5	6	7	8
stochastic	ε, [%]	0.193	0.207	0.223	0.240	0.265	0.296	0.328	0.353
	σ, [MPa]	314.9	328.6	324.9	331.9	335.1	338.9	341.8	345.5
average	ε, [%]	0.183	0.197	0.210	0.224	0.237	0.251	0.265	0.278
	σ, [MPa]	288.5	295.1	300.7	305.2	309.5	310.1	311.1	316.1

2.8.3. Strain hardening zone

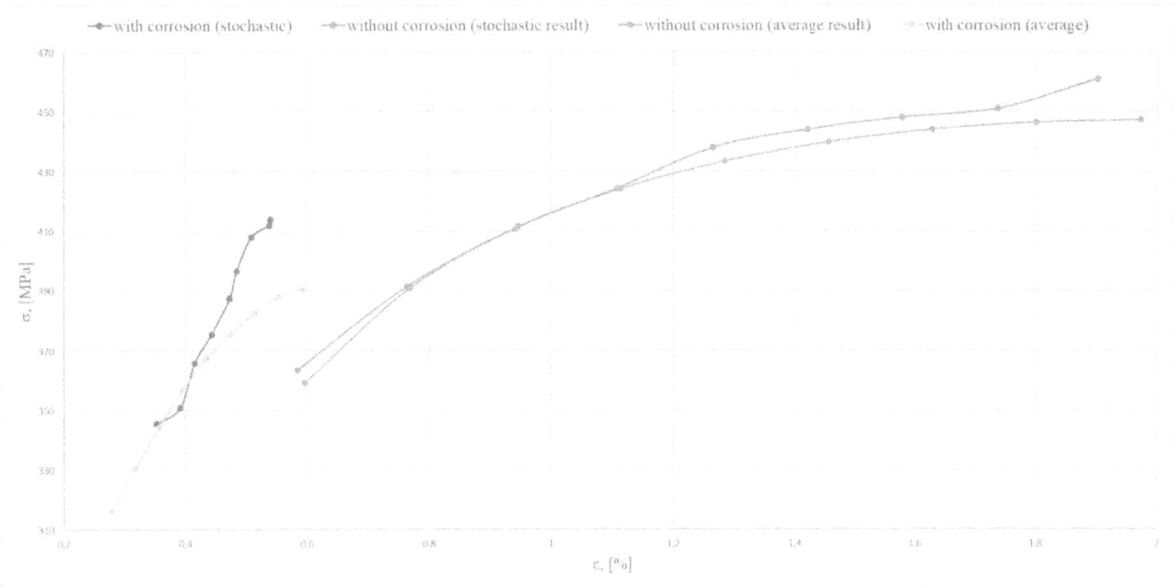

Figure 3.67 Group G strain hardening zone diagrams

In the studied segment of fig. 3.67 and table 3.33, it is found that the strip relative deformation has changed significantly, within the limits of 70%. And the stresses follow a significant decrease of up to 15%. The nonlinear development of mechanical properties is preserved, expressed in a sharp increase in stress relative to the available deformation. It is known that steel has a polycrystalline structure, which means that the volume in the steel contains a large number of small crystals that have a random orientation. The atoms of the metal are arranged in the crystals in a certain way, building a regular spatial lattice. In the lattice there are forces of interaction between the atoms, they form a system that is completely defined for each steel. A clearly expressed change in the crystal lattice of corroded steel has already become a brittle material.

Table 3.33 Data on the main points on which the diagram in fig.3.67

type of result	№ points	1	2	3	4	5	6	7	8
stochastic	ε, [%]	0.392	0.416	0.443	0.473	0.485	0.509	0.538	0.540
	σ, [MPa]	350.9	365.9	375.4	387.4	396.4	407.9	411.9	413.9
average	ε, [%]	0.317	0.357	0.396	0.435	0.475	0.514	0.553	0.592
	σ, [MPa]	330.5	344.2	356.8	367.4	375.4	382.6	387.8	390.6

2.8.4. Necking zone

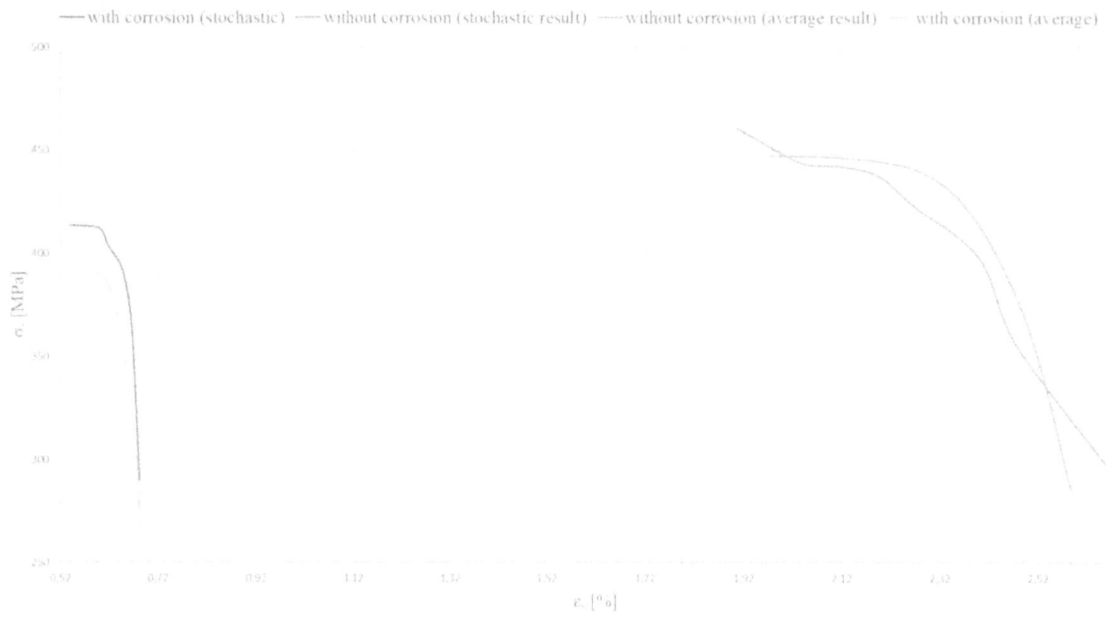

Figure 3.68 Group G necking zone diagrams

In the studied area of fig.3.68 and table 3.34, it is observed that the deformation limit decreases by more than 70%, the rupture stress has a negligible change of 5%. The formation of the neck is observed, i.e. relatively large deformations are localized in a small part of the material and a significant reduction in the cross-sectional area occurs. It is found that there is heterogeneity of the corroded steel. The material begins to deform, all places undergo approximately the same amount of deformation and the greater the local stress, the greater the local reduction in the cross-sectional area, which in turn leads to an even greater concentration of stresses, which leads to instability, which causes the formation of the neck. The destruction occurs at deformation values characteristic of the beginning of the diagram in the strengthening zone before reaching the tensile strength of uncorroded steel.

Table 3.34 Data on the main points on which the diagram in fig.3.68

type of result	№ points	1	2	3	4	5	6	7	8
stochastic	ε, [%]	0.594	0.608	0.617	0.644	0.658	0.666	0.673	0.682
	σ, [MPa]	412.9	409.9	403.8	393.9	379.9	361.9	330.9	289.9
average	ε, [%]	0.604	0.616	0.627	0.639	0.650	0.662	0.674	0.685
	σ, [MPa]	389.1	385.9	378.9	365.2	351.1	330.3	303.2	263.7

2.8.5. Stress-strain diagram

From the studied stress-strain diagram in fig. 3.69 of the corroded material, it can be clearly deduced that corrosion indicates its influence on the mechanical properties of the studied steel, expressed in the following basic facts:

- reduction of steel ductility by over 70%
- there is a change in the structure of the material and it is now classified as a brittle material
- almost no yield point
- slight reduction in strength values up to 5% – 15%
- changes in the internal crystal structure of the steel
- the elastic modulus – decreases by about 35%, has a nonlinear nature and does not obey Hooke's law.

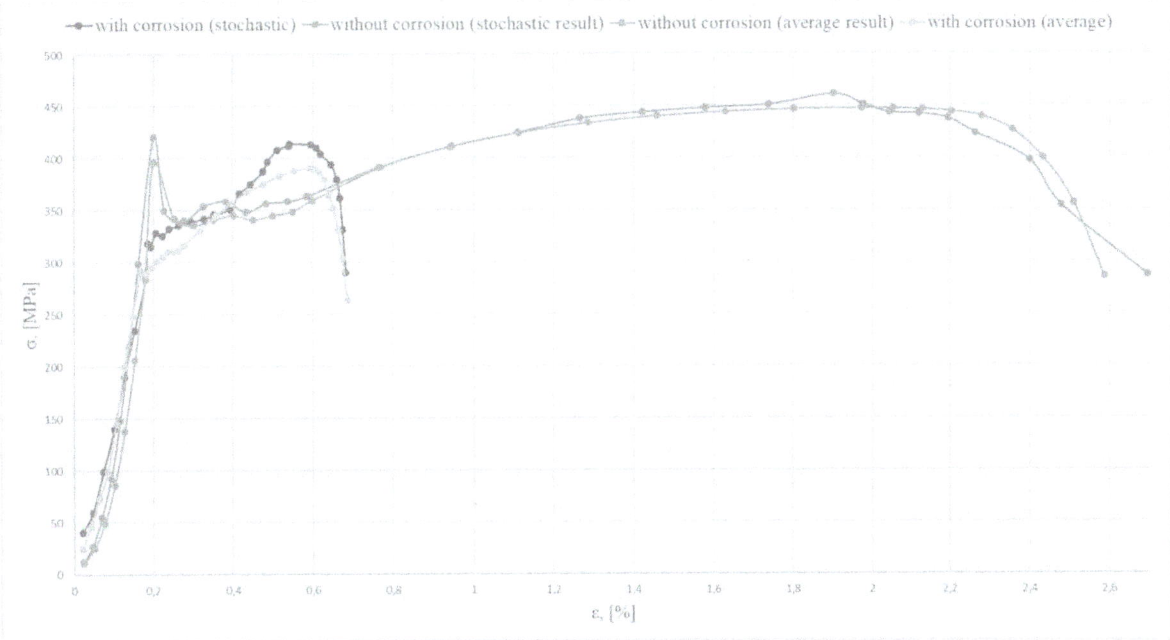

Figure 3.69 Group G stress-strain diagrams

2.9. Study of Group H

Group H consists of 5 corroded steel test specimens with a diameter ranging from 2.00 to 3.00 mm after application of accelerated corrosion. From fig. 3.70 it is established that the change in the diameter change due to accelerated corrosion is nonlinear and does not obey known dependencies. This fact brings us as close as possible to the real conditions of impact, which indicates the cross-section in a corroded steel element. The application of the ISO 9223 standard gives grounds to establish that group

H corresponds to corrosion impact as follows: for corrosion category C1 - 73107 months; C2 - 3801 months; C3 - 1901 months; C4 - 1188 months; C5 - 475 months and CX - 136 months.

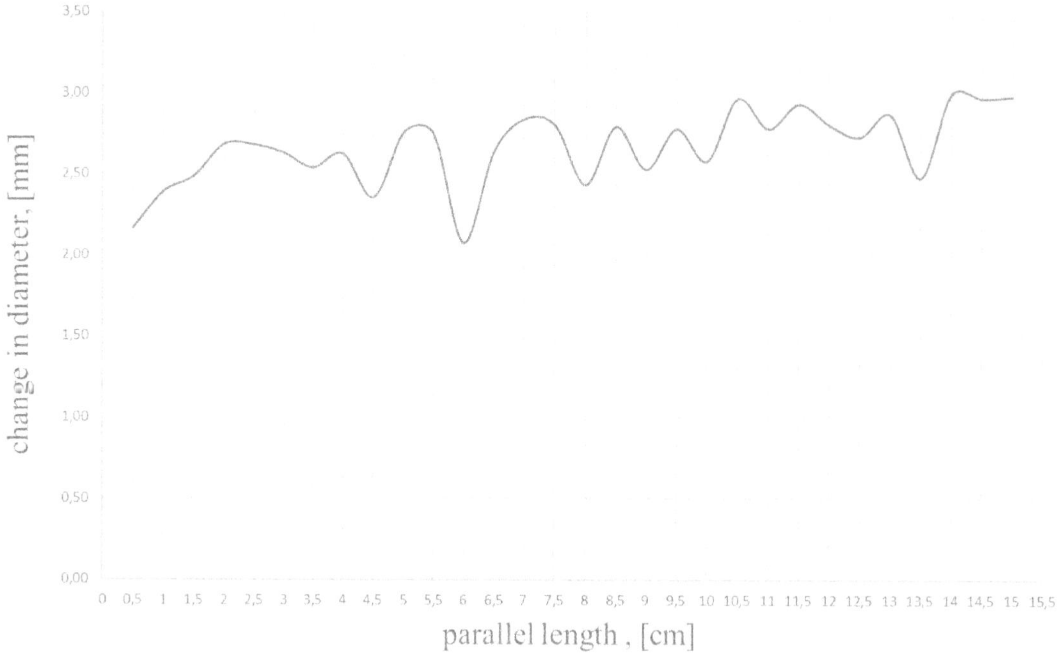

Figure 3.70 Diameter graph after accelerated corrosion in the parallel length – Group H

2.9.1. Elastic zone

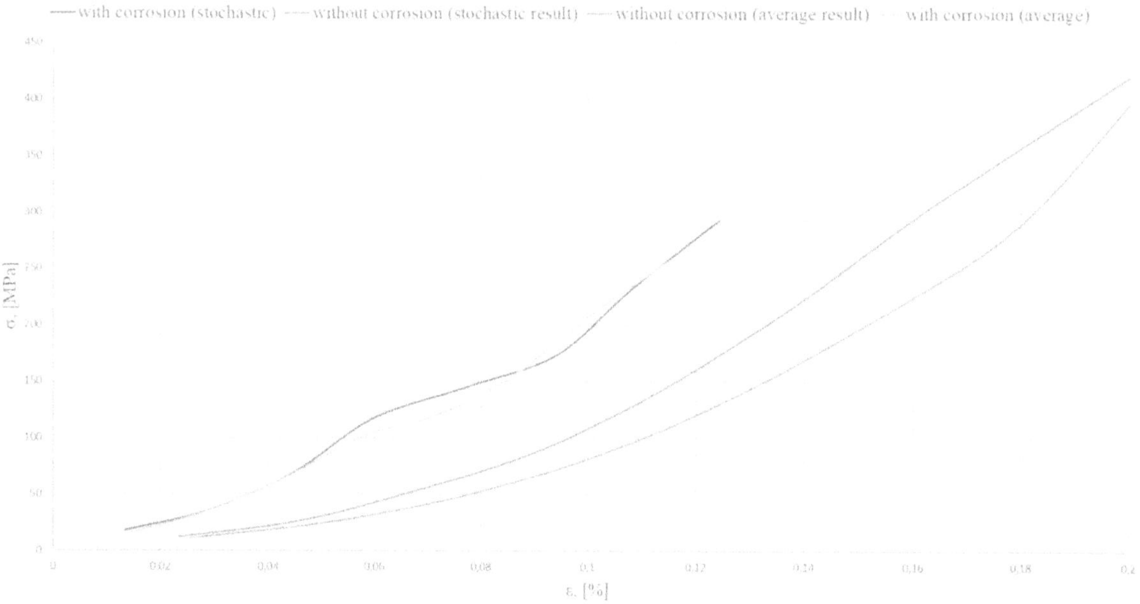

Figure 3.71 Group H elasticity zone diagrams

Table 3.35 Data on the main points on which the diagram in fig.3.71

type of result	№ points	1	2	3	4	5	6	7	8
stochastic	ε, [%]	0.013	0.029	0.045	0.059	0.077	0.094	0.108	0.124
	σ, [MPa]	18.1	35.6	69.7	116.6	144.9	173.9	233.9	292.9
average	ε, [%]	0.014	0.029	0.043	0.058	0.072	0.087	0.101	0.115
	σ, [MPa]	15.4	34.9	64.9	101.3	120.8	156.9	211.2	259.1

The studied area in fig.3.71 and table 3.3, it is observed that for steel without corrosion, the section is rectilinear and follows Hooke's law. For corroded elements, it is established that the diagram begins to change, there is also a reduction in elastic deformation to 40% and a reduction in stresses in the size to 25%. The area begins to lose its elasticity and strength properties. The boundaries within which the relative deformation is striped do not exceed 0.12%. A nonlinear behavior of the modulus of elasticity can be clearly established, although overall a significant reduction is reported.

2.9.2. Yield zone

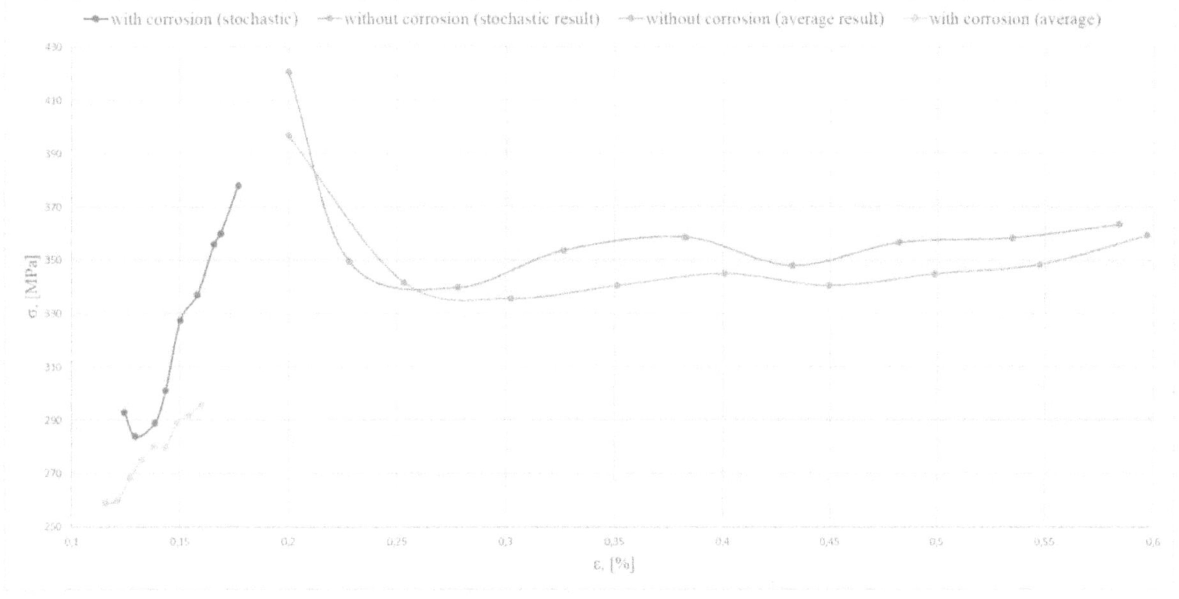

Figure 3.72 Group H yield zone diagrams

The studied area in fig.3.72 and table 3.36 is characterized by the fact that the diagram is distorted, having a clear expression in the vertical direction, but the effect of length in its horizontal part is already starting to be absent. The relative deformation has changed, i.e. the yield point does not exist, and cannot be categorically defined. The stresses show a slight decrease in the range of up to 10%. Circumstances determine that the corroded steel can no longer be classified as a tough material and it has passed into another phase, i.e. it acquires the behavior of a brittle material and means that the destruction should occur even at low values of the deformations. There are plastic deformations in the part, which do not change compared to the uncorroded steel.

Table 3.36 Data on the main points on which the diagram in fig.3.72

type of result	№ points	1	2	3	4	5	6	7	8
stochastic	ε, [%]	0.129	0.138	0.143	0.150	0.158	0.166	0.169	0.177
	σ, [MPa]	283.9	288.9	300.9	327.4	336.9	355.9	359.9	377.9
average	ε, [%]	0.121	0.127	0.132	0.138	0.143	0.149	0.154	0.160
	σ, [MPa]	260.1	268.3	275.2	280.0	279.8	289.0	291.6	295.7

2.9.3. Strain hardening zone

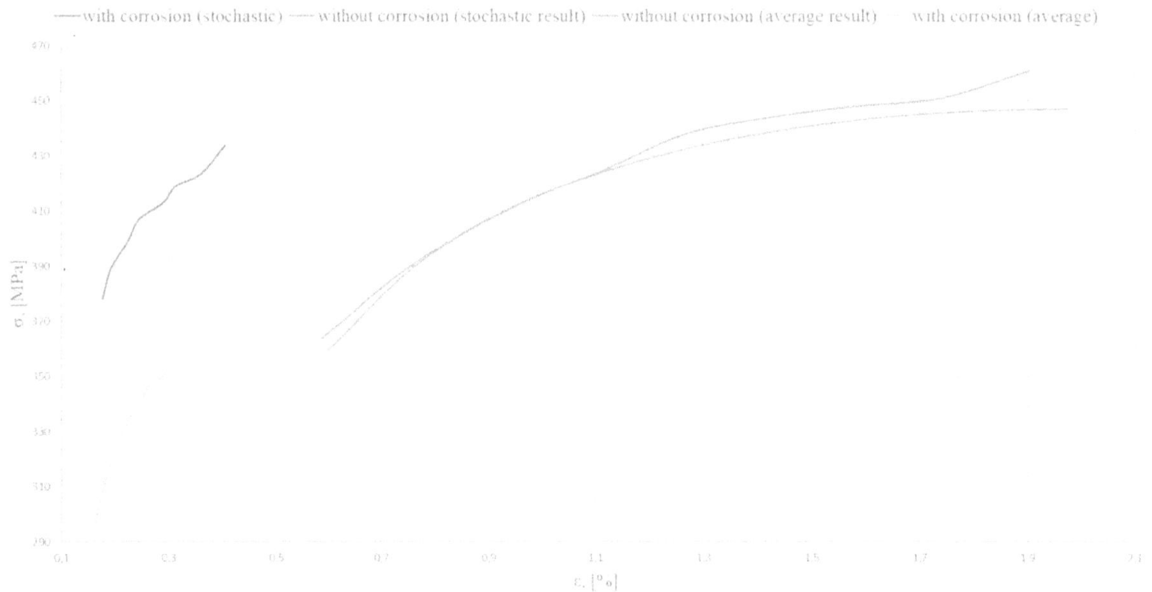

Figure 3.73 Group H strain hardening zone diagrams

In the studied sector of fig.3.73 and table 3.37, it is found that the strip relative deformation has changed significantly, within the limits of 70%. The stresses follow a significant decrease of up to 15%. The nonlinear development of the mechanical properties is preserved, expressed in a sharp increase in the stress relative to the available deformation. It is known that steel has a polycrystalline structure, which means that the volume in the steel contains a large number of small crystals that have a chaotic orientation. The atoms of the metal are arranged in the crystals in a certain way, building a regular spatial lattice. In the lattice there are forces of interaction between the atoms, they form a system that is completely defined for each steel. A clearly expressed change in the crystal lattice of the corroded steel has already become a brittle material.

Table 3.37 Data on the main points on which the diagram in fig.3.73

type of result	№ points	1	2	3	4	5	6	7	8
stochastic	ε, [%]	0.192	0.224	0.245	0.289	0.312	0.347	0.367	0.404
	σ, [MPa]	388.9	398.9	406.9	412.9	418.9	421.9	424.9	433.9
average	ε, [%]	0.178	0.196	0.214	0.232	0.250	0.268	0.285	0.303
	σ, [MPa]	310.1	319.3	329.0	336.5	342.4	347.4	349.5	354.4

2.9.4. Necking zone

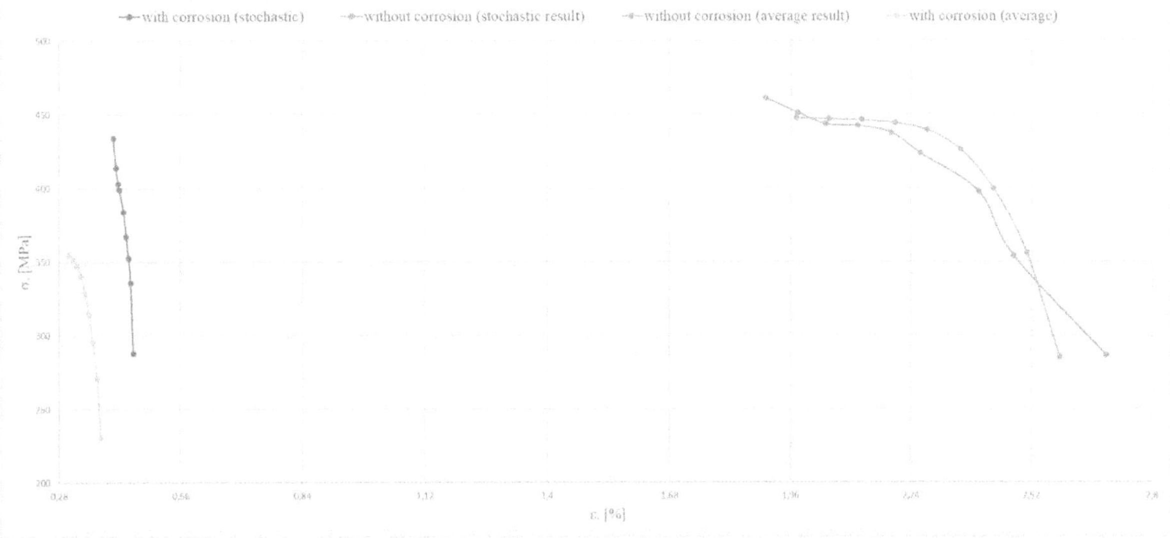

Figure 3.74 Group H necking zone diagrams

In the studied sector in fig.3.74 and table 3.38, the result shows that the deformation limit decreases by more than 75%, the failure stress has a negligible change of 5%. The formation of a neck is observed, i.e. relatively large deformations are localized in a small part of the material and a significant reduction in the cross-sectional area occurs. It is found that there is heterogeneity of the corroded steel. The material begins to deform, all places undergo approximately the same deformation. The greater the local stress, the greater the local reduction in the cross-sectional area, which in turn leads to an even greater concentration of stresses, this leads to instability, which causes the formation of a neck. The failure occurs at deformation values characteristic of the end of the diagram in the yield zone of non-corroded steel.

Table 3.38 Data on the main points on which the diagram in fig.3.74

type of result	№ points	1	2	3	4	5	6	7	8
stochastic	ε, [%]	0.410	0.415	0.418	0.428	0.434	0.439	0.444	0.450
	σ, [MPa]	413.9	402.9	398.9	383.9	366.9	352.4	335.7	287.9
average	ε, [%]	0.312	0.321	0.330	0.339	0.348	0.357	0.366	0.375
	σ, [MPa]	351.1	347.6	340.3	329.3	314.4	295.6	270.9	231.0

2.9.5. Stress-strain diagram

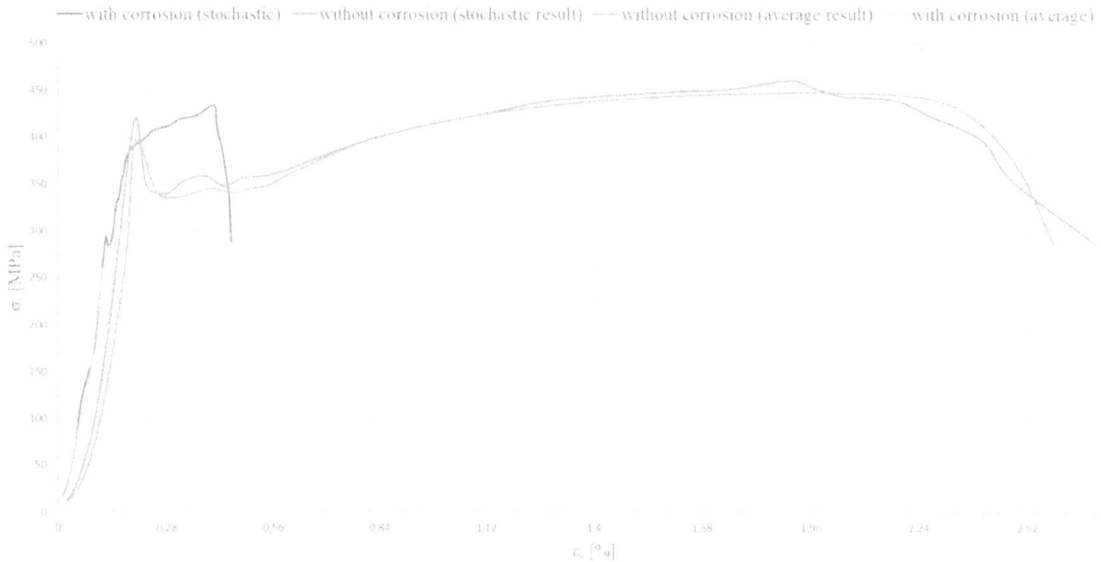

Figure 3.75 Group H stress-strain diagrams

From the studied stress-strain diagram in fig. 3.75 of the corroded material, it is clearly seen that corrosion indicates its influence on the mechanical properties of the studied steel, expressed in the following basic facts:

- reduction of steel ductility by over 70%
- there is a change in the structure of the material and it is now classified as a brittle material
- missing yield zone
- slight reduction in strength values up to 5% – 15%
- changes the internal crystal structure of the steel

- the modulus of elasticity – decreases by about 35%, has a nonlinear nature and does not obey Hooke's

2.10. Study of Group I

Group I consists of 3 corroded steel test specimens with a diameter varying after application of accelerated corrosion from 1.00 to 2.00 mm. fig. 3.76 shows the change in diameter in the parallel length and the change due to accelerated corrosion is nonlinear and does not obey known dependencies, bringing us as close as possible to the real conditions of impact. Using the ISO 9223 standard, it is reasonable to establish that group I responds to corrosion impact as follows: C1 - 77262 months; C2 - 4018 months; C3 - 2009 months; C4 - 1256 months; C5 - 502 months and CX - 143 months.

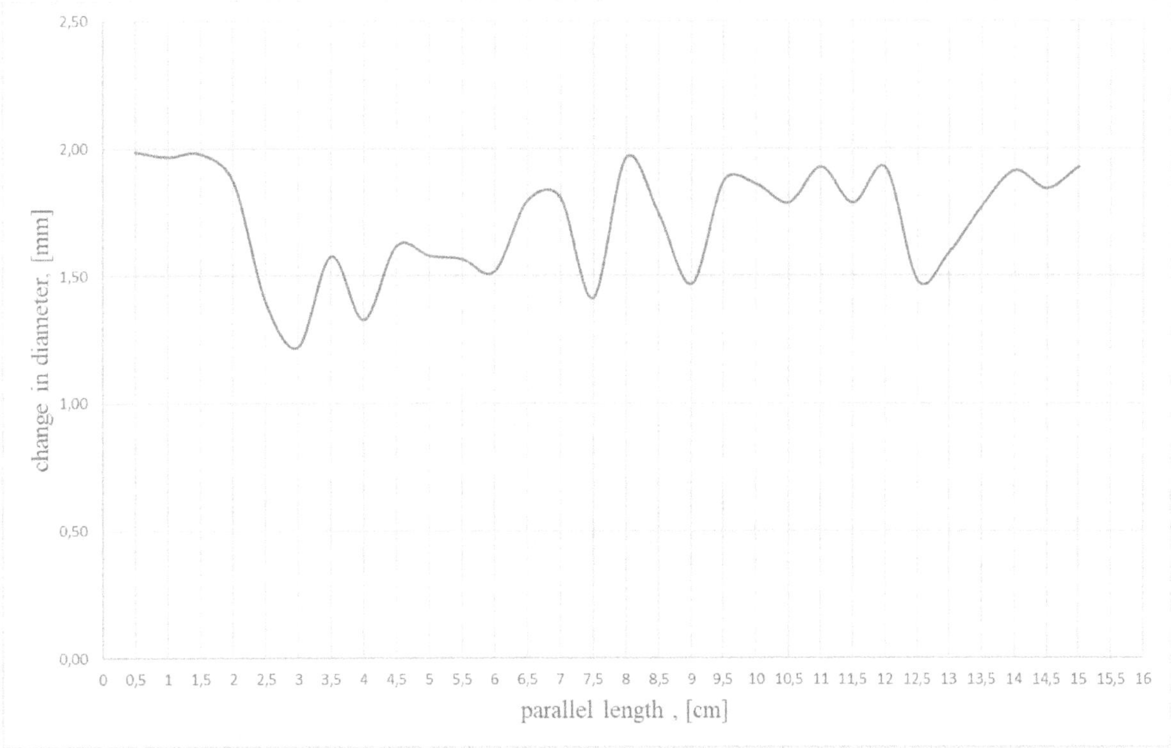

Figure 3.76 Diameter graph after accelerated corrosion in the parallel length – Group I

2.10.1. Elastic zone

The studied section in fig.3.77 and table 3.39 is characterized by the fact that for steel without corrosion, the section is straight and follows Hooke's law. For corroded elements, it is found that the diagram begins to change, as there is already a reduction in elastic deformation to 40% and leads to a significant reduction in stresses in the size of up to 25%. In general, the area begins to lose its elasticity and strength properties. The boundaries within which the relative deformation is located do not exceed

0.12%. It can be clearly established that the modulus of elasticity behaves non-linearly, and in general it reports a significant decrease.

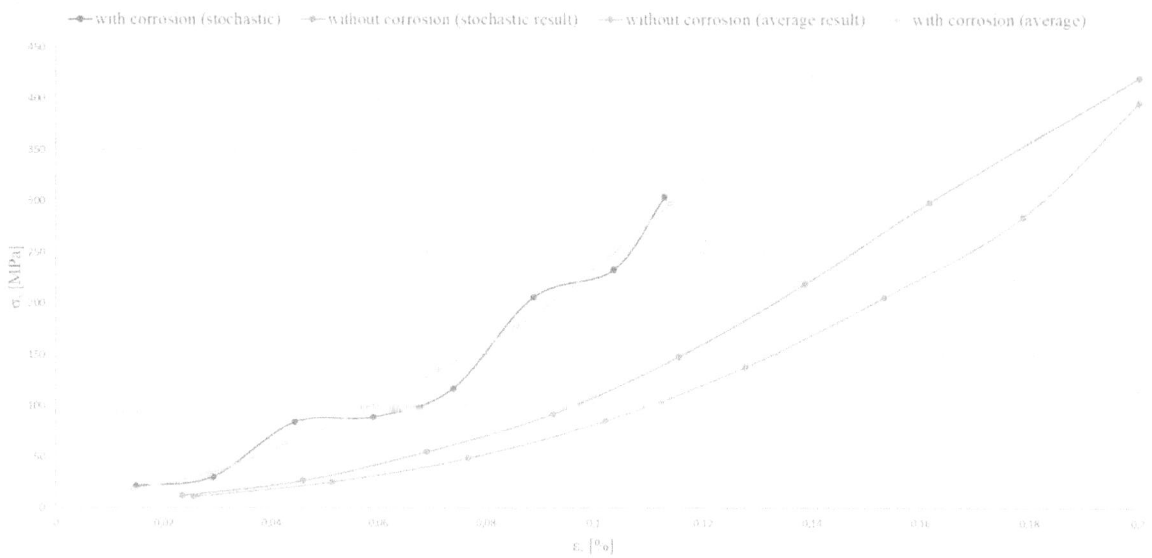

Figure 3.77 Group I elasticity zone diagrams

Table 3.39 Data on the main points on which the diagram in fig.3.77

type of result	№ points	1	2	3	4	5	6	7	8
stochastic	ε, [%]	0.015	0.029	0.044	0.059	0.074	0.089	0.104	0.113
	σ, [MPa]	21.3	30.3	83.8	88.9	116.9	205.9	232.9	303.9
average	ε, [%]	0.014	0.028	0.043	0.057	0.071	0.085	0.100	0.114
	σ, [MPa]	18.1	34.1	61.9	95.7	134.8	176.6	232.2	298.0

2.10.2. Yield zone

In the studied area of fig.3.78 and table 3.40, it is established that the diagram is distorted, having a clear expression in the vertical direction, but the effect along the length in its horizontal part has already disappeared. It is established that the relative deformation has changed, i.e. the yielding area does not exist, and cannot be categorically defined, but the stresses have a slight decrease in the range of up to 10%. Circumstances determine that the corroded steel can no longer be classified as a tough material and it has passed into another phase, i.e. acquires the behavior of a brittle material and means that the destruction should occur even at low values of the deformations. There are residual deformations in the part, which do not change compared to the non-corroded steel.

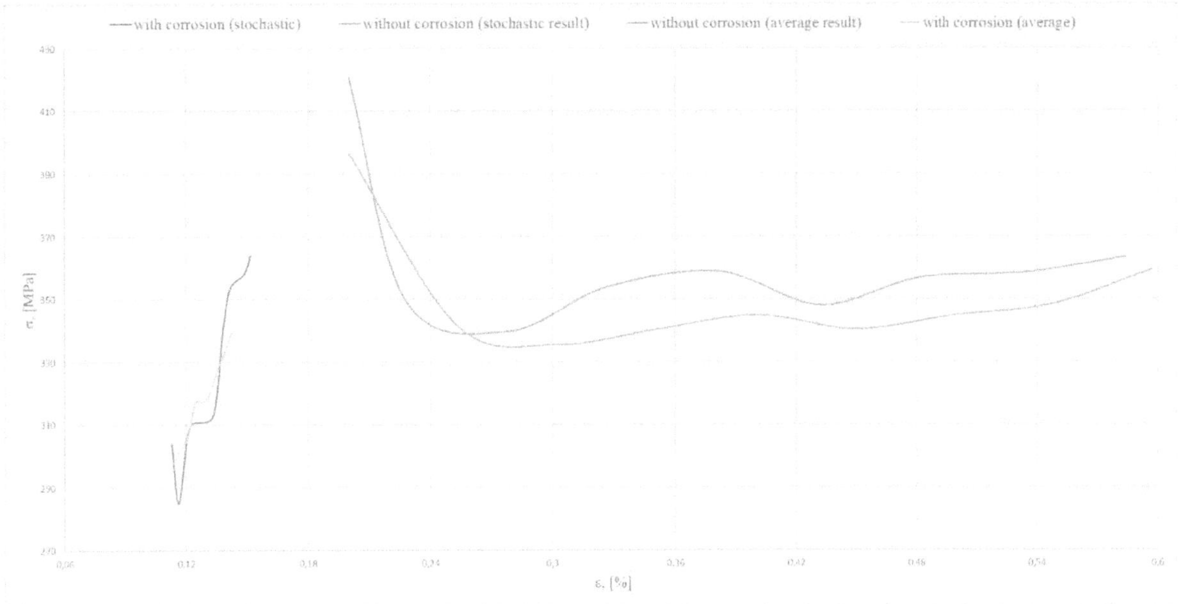

Figure 3.78 Group I yield zone diagrams

Table 3.40 Data on the main points on which the diagram in fig.3.78

type of result	№ points	1	2	3	4	5	6	7	8
stochastic	ε, [%]	0.116	0.120	0.123	0.133	0.138	0.142	0.148	0.152
	σ, [MPa]	284.9	305.9	310.4	312.9	340.4	353.4	357.9	363.9
average	ε, [%]	0.117	0.121	0.125	0.128	0.132	0.135	0.139	0.142
	σ, [MPa]	302.3	306.3	317.1	317.1	319.5	327.3	332.4	339.4

2.10.3. Strain hardening zone

Table 3.41 Data on the main points on which the diagram in fig.3.79

type of result	№ points	1	2	3	4	5	6	7	8
stochastic	ε, [%]	0.163	0.174	0.181	0.188	0.195	0.202	0.214	0.229
	σ, [MPa]	384.9	401.9	418.9	421.9	425.9	429.9	431.9	451.9
average	ε, [%]	0.154	0.166	0.178	0.190	0.201	0.213	0.225	0.237
	σ, [MPa]	357.7	371.3	385.2	396.4	405.2	412.7	417.4	421.0

In the investigated area of fig.3.79 and table 3.41, it is found that the strip relative deformation has changed significantly, within the limits of 80%. And the stresses follow a significant decrease of up to 15%.

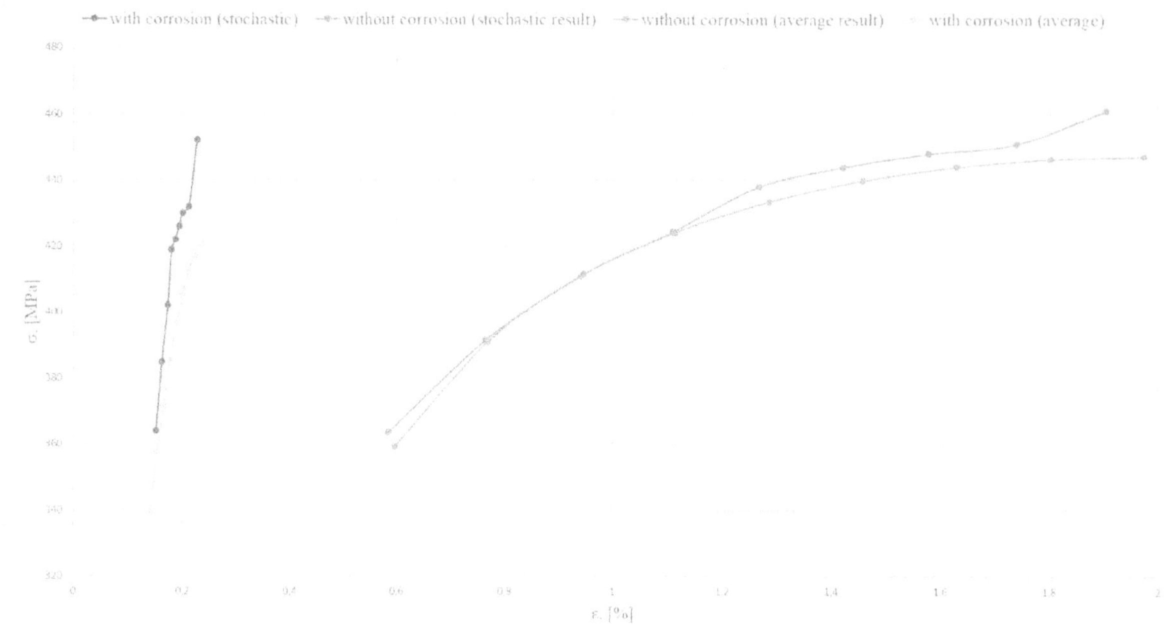

Figure 3.79 Group I strain hardening zone diagrams

The nonlinear development of mechanical properties is preserved, expressed in a sharp increase in stress relative to the available deformation. It is known that steel has a polycrystalline structure, which means that the volume in the steel contains a large number of small crystals that have a chaotic orientation. The atoms of the metal are arranged in the crystals in a certain way, building a regular spatial lattice. In the lattice there are forces of interaction between the atoms, they form a system that is completely defined for each steel. A clearly expressed change in the crystal lattice of corroded steel has already become a brittle material.

2.10.4. Necking zone

Table 3.42 Data on the main points on which the diagram in fig.3.80

type of result	№ points	1	2	3	4	5	6	7	8
stochastic	ε, [%]	0.235	0.239	0.244	0.252	0.256	0.326	0.341	0.344
	σ, [MPa]	440.9	427.9	421.8	413.9	404.9	391.9	371.9	322.9
average	ε, [%]	0.243	0.249	0.256	0.262	0.268	0.275	0.281	0.287
	σ, [MPa]	418.4	415.2	409.8	402.4	388.1	367.6	340.6	295.3

In the studied sector of fig.3.80 and table 3.42, it is determined that the deformation limit decreases by more than 75%, the failure stress has a negligible change of 5%. The formation of the neck is observed, i.e. relatively large deformations are localized in a small part of the material and a significant reduction in the cross-sectional area occurs. It is found that there is heterogeneity of the corroded steel. As the material begins to deform, all places undergo approximately the same amount of deformation and the greater the local stress, the greater the local reduction in the cross-sectional area, which in turn leads to an even greater concentration of stresses, which leads to instability, which causes the formation of the neck. The failure occurs at deformation values characteristic of the end of the diagram in the elastic zone of non-corroded steel.

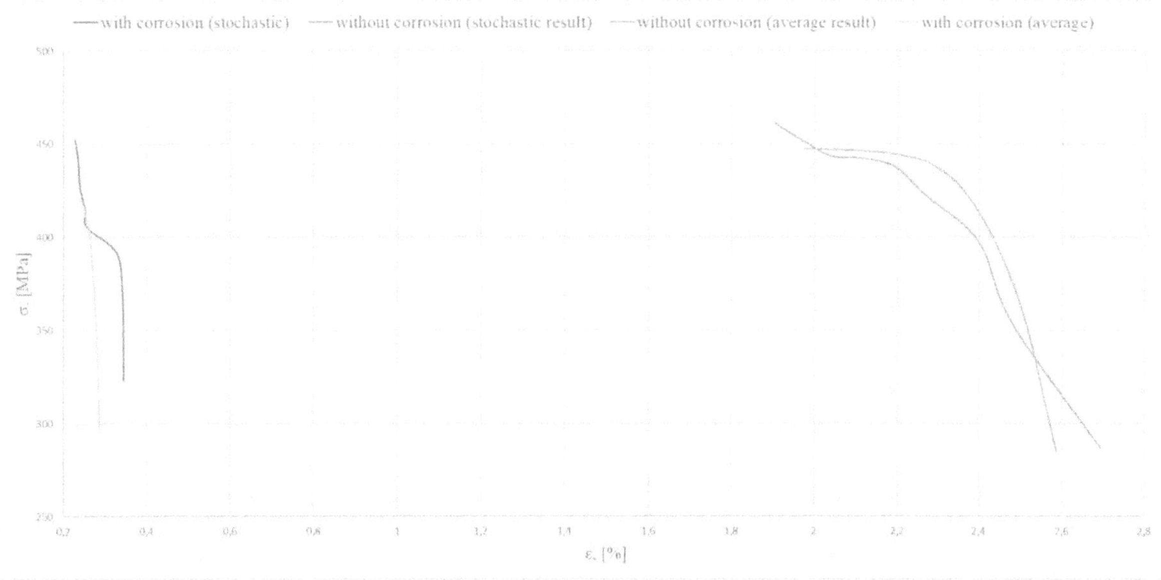

Figure 3.80 Group I necking zone diagrams

2.10.5 Stress-strain diagram

The stress-strain diagram in fig. 3.81 shows that corrosion has an influence on the mechanical properties of the steel under study, expressed in the following basic facts:

- reduction of steel ductility by over 75%
- there is a change in the structure of the material and it is now classified as brittle
- missing yield zone
- slight reduction in strength values up to 5% – 15%
- changes the internal crystal structure of the steel

- modulus of elasticity – decreases by about 35%, has a nonlinear nature and does not obey Hooke's law

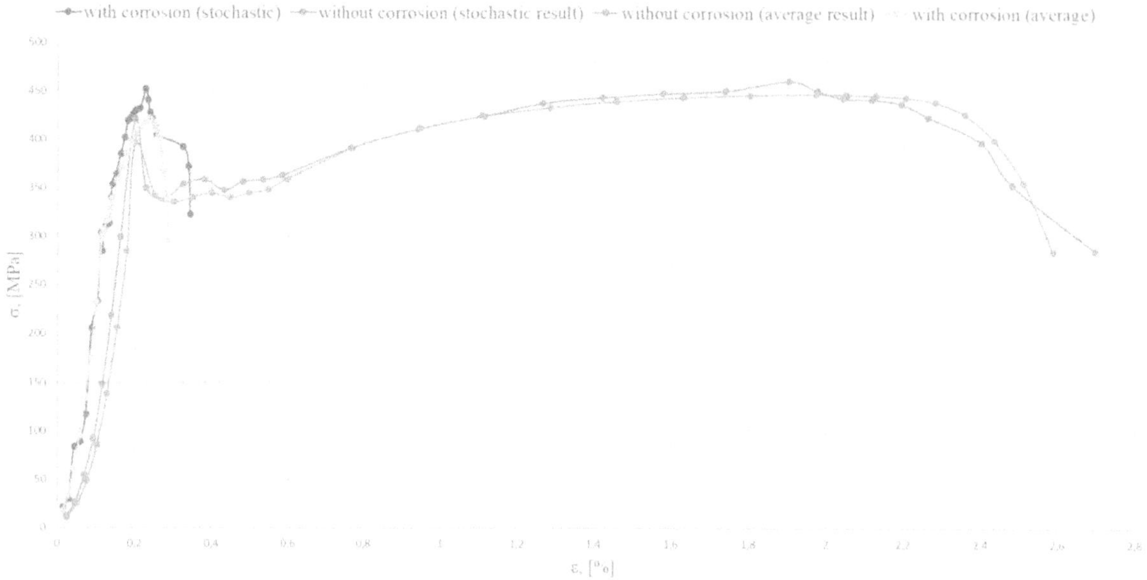

Figure 3.81 Group I stress-strain diagrams

3. Analysis of results

3.1 Elastic zone

At the beginning of corrosion development in steel samples, a change in the ductility of the material is observed, but it follows Hooke's law. The change in values is up to 12%. As corrosion progresses, its impact begins to affect the zone, starting to increase the elastic deformation to 20%. With continued development of the corrosion effect, a moment is reached when the elastic deformation is increased, but at the expense of the strength properties, which decrease to 15%. The more corrosion develops in the steel element, the more the ductility in the zone begins to decrease and at a certain point it has already lost its increased elasticity, respectively, the drop in stresses remains the same. With corrosion over 40% of the weight, with continued corrosion influence it begins to lead to a change in the dependence of deformation, i.e. it is no longer a linear deformation and does not follow Hooke's law, the ductility has decreased dramatically by more than 50%.

3.2 Yield zone

With developed corrosion up to 15% of the weight, it is established that the relative deformation has not changed, the ductility also, the yield point is clearly expressed, the values mark a slight decrease of up to 5%. The behavior of an elastoplastic material is observed. As the corrosion progresses, it is found that the ductility begins to decrease to 25%, there is still a yield point, although the material cannot be classified as tough. With increasing corrosion development, the yield point disappears, the ductility decreases significantly over 50%, the material acquires the behavior of a brittle material. A clearly expressed vertical section in the zone is established.

3.3 Strain hardening zone

When corrosion begins to develop, no significant changes are observed in the zone. Ductility has been preserved, with a slight decrease in the strength characteristics of the material of up to 5%. The polycrystalline structure of the steel contains a crystal lattice and the existing forces interact within it, but the sliding cannot grow indefinitely, the structure has changed as a result of the plastic deformations that have occurred, which is established by the weaker slope of the diagram. With the continued impact of corrosion, ductility and relative deformation begin to show a sharp decrease to 50%, and strength retains its trend of increase, i.e. a "corrosion strengthening" effect of 10% appears. As the corrosion effect increases, the research shows that there is a clear change in the crystal lattice of the steel, the transformation from an elastoplastic material to a brittle material has already been completed, the ductility has significantly decreased, as has the relative deformation of 70%.

3.4 Necking zone

With the onset of corrosion, the study in the area shows that the ductility decreases directly by up to 15%. Although the values of the fracture strength are found to be within significantly close limits, the loss of ductility is significant from the very beginning. With increasing corrosion, the ductility begins to decrease, reaching values of up to 40%. The more corrosion enters the steel element, the more its ductility decreases until it acquires ductile behavior characteristic of brittle materials. Although the fracture strength does not change within sensitive limits.

3.5 Stress-strain diagram

The basic definition of ductility is the ability of a material to undergo significant plastic deformation before breaking, a measure that can be expressed as a percentage during the processes of elongation or reduction in area, in a tensile test. Ductility is a fundamental property of steel, as materials that crack or break under stress cannot be manipulated using metalworking processes such as impact, rolling, drawing or extruding. High degrees of ductility are found in steel elements, as well as in other materials in metals, leading to the common perception that metals are generally ductile. In metallic bonds, the valence electrons of the shell are delocalized and shared among many atoms. The delocalized electrons allow the metal atoms to slide past each other without being subjected to strong repulsive forces that would cause other materials to break. Elasticity can be quantified by fracture strain, which is the stress at which a specimen fails in a uniaxial tensile test. Another commonly used measure is the reduction in fracture area. The elasticity of structural steel varies depending on the alloying elements. Increasing carbon levels reduce the ductility of the steel, i.e. it becomes brittle steel. At corrosion up to 20%, the change in elasticity is small, but as corrosion progresses in steel load-bearing elements, the change in ductility becomes significant. In terms of strength, there is a change of up to 10%, which is due to the corrosive effect. As corrosion progresses, the corresponding ductility decreases.

4 Microstructural study

After the tensile test, a random element is selected for microstructural examination in order to trace the destructive processes in the surface layer during corrosion. In connection with the determination of the interdependence of all indicators that affect the tensile strength of steels during corrosion, a need arises for a systematic approach to the problems [49]. From the point of view of statistical data, and the prediction and monitoring of corrosion for the possible development of processes in real operating conditions are the basis for determining the required level of structural reliability of energy-loaded products and structures that operate in conditions of complex and stressed state [49].

It is obvious that the adoption of strength calculation methods based on the principles of maximum load or allowable stresses does not necessarily exhaust all aspects of the destruction [49]. Therefore, a scanning electron microscope observation was carried out with a magnification of up to 5000 times of the formed corrosion fig. 3.82. On fig. 3.82(b) shows detached grains of corrosion, which does not affect the mechanical properties of the steel. The phenomena are associated with the destruction of materials, with unstable growth of the initial moment of microcrack formation and stress concentration [49]. A star-shaped structure is observed at a magnification of 1000, 2500 and 5000 in Fig. 3.82(a), Fig. 3.83. The resulting structure from Fig. 3.84 can be explained by the tensile load on the test specimens and the uneven stress applied during the corrosion process on the test specimens – 105 pieces.

The influence of tensile stress on the metal combined with corrosion failure is observed in fig. 3.84.

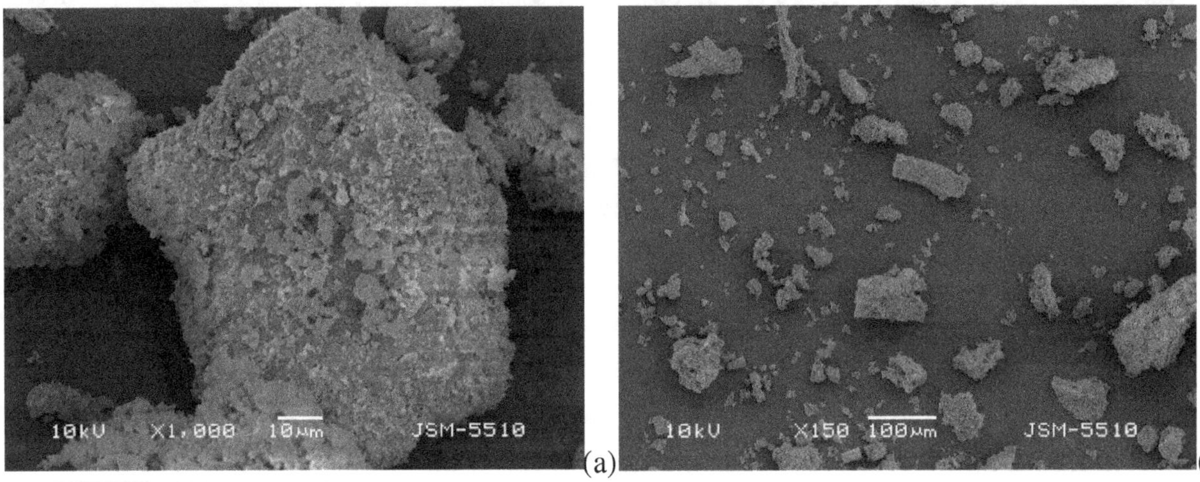

Figure 3.82 (a) Magnification 1000 times [49]; (b) Magnification 150 times

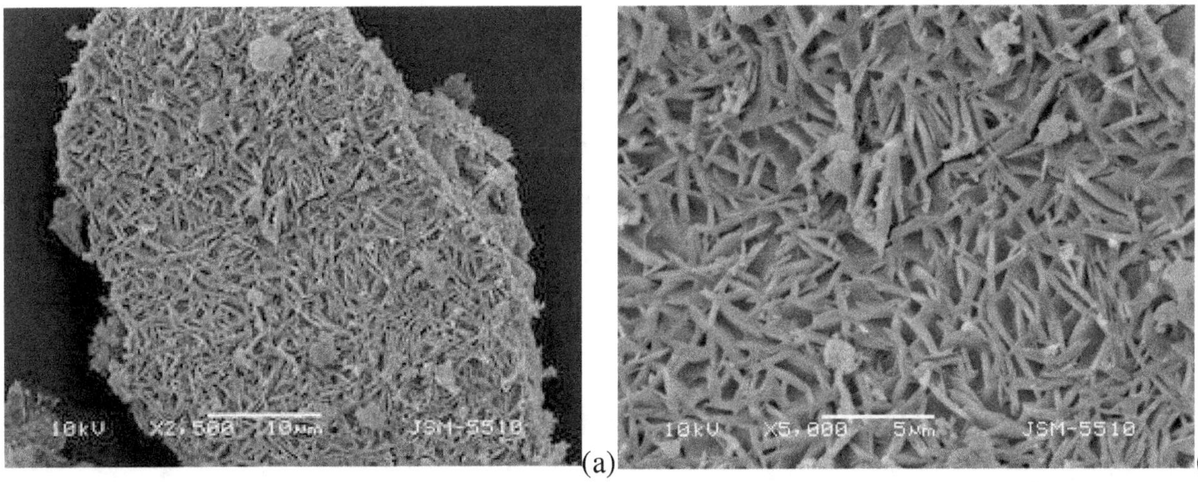

Figure 3.83 (a) Magnification 2500 times; (b) Magnification 5000 times [49]

The boundary between the base metal and the corrosion coating is clearly observed. All still have a strong connection, which is evident from the photographs with a magnification of 1000, a segment of the boundary is shown in Fig. 3.84(b) to observe its change [49]. It is necessary in the cases shown in Fig. 3.84(a), to improve the energy reserve, which affects the nature of the destruction and the strength of the system [49]. The magnitude of the energy reserve in a stressed system is determined by the change in the loading forces when determining the results of the destruction [49]. Over-storage of energy leads to reduced relaxation forces, and the failures have an avalanche-like character. Increasing the energy reserve is dangerous for materials that are sensitive to the influence of defects, characteristic of complex systems [49]. With a decrease in the degree of relaxation, the rate of external deformation force increases.

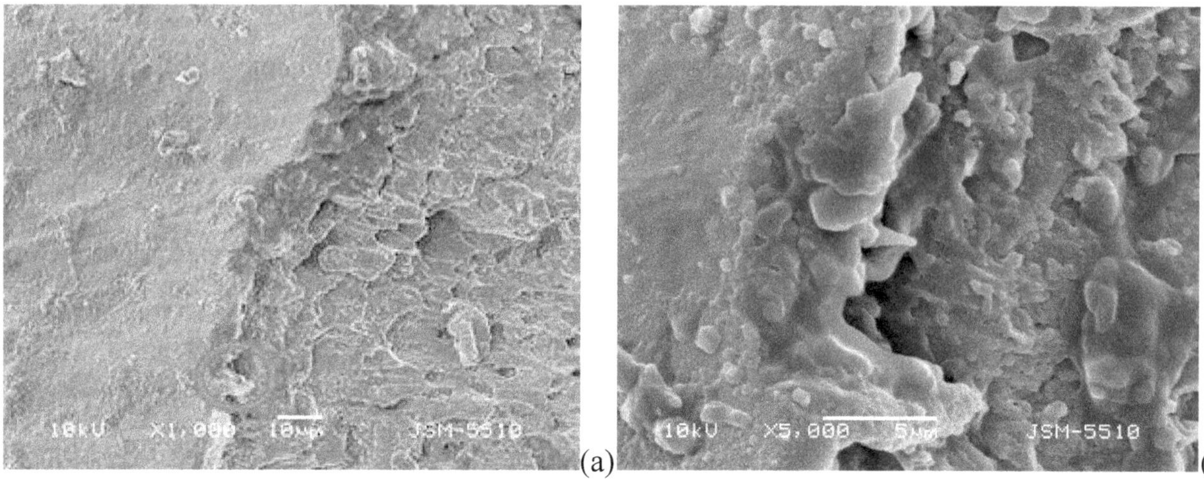

(a) (b)

Figure 3.84 (a) Corrosion limit, magnification 1000 times [49]; (b) Segment of the border (red circled), magnification 5000 times [49]

In the formation of cracks in steel with corrosion, one should look for the relationship between the energy characteristics that underlie the energy theory of destruction, first proposed by Alan Griffith [102]. It does not take into account the occurrence of cracks, but only the conditions for the occurrence of a critical state of destruction of the body [49]. It does not take into account the time and kinetics of the process, both before and after the critical state [49]. For the kinetics of the process, it is necessary to assume a dependence of the surface tension on time and to reflect the processes that accompany the destruction of the bodies [49]. From fig. 3.85 it is established that the uneven penetration of corrosion through oxidation of iron into the main steel alloy.

(a) (b)

Figure 3.85 Optical microscope (a) Magnification 1000 times; (b) Magnification 500 times [49]

This leads to different content, quality and quantity of oxides [49] – Fe_2O_5 , FeO, FeO_2 and FexOx. Part of the corrosion cannot be cleaned, since oxidation processes have started in the microstructure of the steel and the formation of microcracks in the cross section has started [49]. Fig. 3.86 shows the structure of the metal under tensile load and the state of corrosion [49]. This is necessary to monitor the state of the steel structure and prove that stress arises in the corrosion layer and corrosion absorbs part of the stress [49]. The fact confirms that stress exists inside the corrosion [49]. In addition to the surface, it has also entered the boundary with the base of the metal. It was necessary to do this in order to monitor the structural changes.

Figure 3.86 (a) Magnification 1100 times [49]; (b) Magnification 2500 times

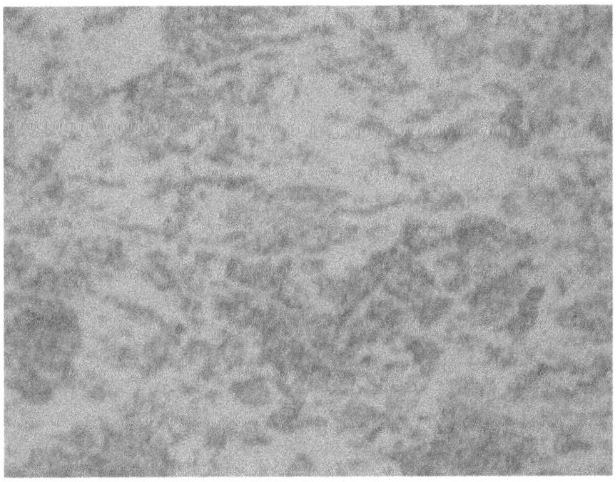

Figure 3.87 Metallographic photograph

The metallographic analysis of steel S355JR, shown in fig. 3.87, establishes the microstructure. It consists of ferrite and pearlite, typical of low-carbon steels. The results of the analysis show that in corrosion failure, the brittle mechanism of failure predominates, characterized by the propagation of cracks along certain crystal planes. In the boundary areas, traces of ulcers with a complex micro relief are noticeable, occurring at the boundary both in and between the grains of pearlite and ferrite. This is probably an incubation period and a reason for increasing the rate of crack growth.

5 Summary

A modern accelerated corrosion methodology is applied, which allows for effective diagnosis and assessment of the degree of corrosion development per unit of time.

The chemical method for cleaning corroded test specimens from corrosion products has been modernized.

By using the stochastic method, it is proven that the most reliable results are obtained in time in view of the probabilistic event of the calculations.

A probabilistic model has been created using modern mathematical apparatus.

Stages of the working diagram without corrosion have been studied and the working diagram of the material steel S355JR has been summarized, reliable results have been obtained certifying 84-87% accuracy

105 test specimens have been tested and the obtained parts of the working diagram of the material have been processed using mathematical models, they are sufficiently reliable over 85% probability and can be used in practice.

Using scanning electron microscopy, the change in the resulting corrosion, S355JR steel and the corrosion/steel interface was monitored.

Chapter IV. Prediction of corrosion impact

IV.A. Theoretical modeling

1. Introduction

The problem of corrosion in steel elements and structures in construction is shown in fig. 4.1. Methods are being studied for predicting the corrosion effect, which has a negative impact on the main characteristics - a decrease in geometric characteristics, the appearance of surface defects, structural changes in the material and the strength-deformation properties of steel [67, 68, 72, 78, 87, 88, 93, 96], which also affects the parameters of technical facilities and systems.

Figure 4.1 Corroded load-bearing steel elements in structures [110]

2. Model of the corrosion process

Anti-corrosion protection of steel structures and elements is a practically solvable task. Under operational conditions, anti-corrosion protection measures have been developed - paint coatings and systems [68, 93], which, however, significantly increases operating costs, as with an increase in the number of coatings applied over time, their reliability in terms of the structure decreases.

The experimental data for the study of metal structures and elements operated in aggressive corrosive environments and after processing can be established the following formula for satisfying the corrosion effect, describing it with an exponential function (fig.4.1):

$$\Pi = A.m^{n} \tag{4.1}$$

where: Π are corrosion losses, mm ; A and m are constant parameters (depending on the environment in which the element is located), n is the operating time in years.

The n index in most of the studied cases of steel is close to 1/3, and the age coefficient A varies from 0.08 for slightly aggressive environments to 0.45 for highly aggressive environments [78]. Considering the regularity, as well as the specific factors of the elements such as the probability of corrosion and the unevenness of fractures along the perimeter of the cross-section of the elements, it is possible to predict the variation of the bearing capacity of metal structures during operation, which is especially necessary when reconstructing building composite structures and assessing the suitability of structures for further operation.

3. Forecasting the development of the corrosion process

3.1. Forecast

Forecast is the probabilistic assessment of the state of a given object and is a process and/or phenomenon at a given time in the future, for example, in terms of corrosion. In relation to the problem under consideration, this is a prediction of the formed indications of the development of the process. The model obtained by the methods described and used for forecasting is called a forecast. It must meet the following requirements:

- to correspond to the intended completeness, adaptability and evolution, to allow the introduction of changes and additions to consistently approach the required accuracy;
- be abstract enough to allow for varying the number of necessary factors and to ensure the reliability of the results obtained;
- have time constraints to achieve quick results through short-term forecasting and exclude large machine time costs through long-term forecasting or evaluation of forecast results;
- be oriented towards implementation using known technical means and be built on the basis of established terminology;
- to provide useful information about the process in relation to the research task;
- to enable verification of the truthfulness of real reporting of actual corrosion processes

The principles and rules uniting and separating the common properties of models are set out in the special literature [103].

3.2. Types of models

Depending on the number of variables involved in the description of the model at the analysis stage, the predictive objects are divided on a quantitative scale into [103, 104]:

- sub-local (1 ... 3 factors);

- sub-global (15... 35 factors);
- global (with the number of significant variables 36 ... 100);
- superglobal (more than 100 factors); [103].

Corrosion processes and their prediction mainly concern the first three classes. In particular, atmospheric corrosion is often the subject of such prediction [78].

In addition to the quantitative scale, the prediction method is significantly influenced by the complexity of the object:

- super simple - these are objects that do not have significant interrelationships between variables. For them, predictions can be constructed through a sequential analysis of the independent variables Xt (factors) that make up the description of the objects at any scale;
- simple - in the description of which there is a pair of relationships between the variables. In these cases, forecasts are created using pair regression models, simple expert methods for assessing the degree and nature of relationships between variables;
- complex – in which, with an appropriate description, it is necessary to take into account the relationships and joint influences of several significant factors - from three or more, while at the same time the main influencing factors can be clearly outlined. As a mathematical apparatus in this case, methods for multiple regression and correlation analysis, expert tables for assessing mutual influence and preferences are most often used;
- super complex - objects, in the description of which it is necessary to take into account the interrelationships between all variables. The main methods for studying them are multiple, correlation, factor and variance analysis, while constantly taking into account the reliability of the forecast [104, 106].
- According to the factors of development of the forecast, we can divide them into two main groups:
- discrete - a regular component (trend) that changes periodically at certain points in time, as a regular component is understood as a description of the process that is free from a random component - noise or other random factors not inherent to the environment or element);
- aperiodic, describing the regular component in the form of an aperiodic continuous function of time.

Corrosion processes are mainly due to aperiodic factors and less often to discrete ones.

3.3. Modeling

In the process of operation of structures, facilities and equipment, it is often necessary to solve tasks related to the assessment of the actual state and forecasting its change in order to promptly take measures to prevent the moment of failure of the structure and establish the real reserve of technical suitability.

If the determination of the current corrosion state basically leads through its results to an immediate assessment of the corrosion effects and established consequences, then forecasting based on experimental data also implies the availability of opportunities for developing hypotheses allowing to outline the course of development of corrosion processes and the change in the technical state and behavior of the object in the future. In these cases, physical and mathematical modeling is used.

Physical modeling for assessing the state of objects and forecasting the development of corrosion processes has limited capabilities in making decisions in individual cases [85, 86, 98, 99, 103, 106].

Mathematical modeling offers significantly greater, universal, and practically inexhaustible possibilities [103, 106].

The general principles of the development of mathematical models and their refinement for the assessment of the residual service life of structures, facilities and equipment are presented in groups in the presentation below.

Process modeling implies the creation of a logical, analytical, graphical or any other accurate description, corresponding to reality and allowing to analyze and evaluate the dynamics of its development depending on the specific conditions. The type of description in most cases is determined by the purpose of the developed model and the availability of statistical material and/or research, making it possible to assess the degree of approximation of the model to the actually occurring processes. In the case of models reflecting the dependence of the corrosion process on time, occurring in stationary conditions, it is necessary to have statistical material available for the depth of corrosion - lк, which can be expressed analytically as a function of time (form. 4.2) and thus to obtain a simplified model for the solution of the problem of the type

$$l_k = v_0.\tau \tag{4.2}$$

Where: v_0 is the initial corrosion rate; τ is a time.

If it is necessary to assess the change in the state of the structure or object by the degree of its corroded elements, occurring in non-stationary operating conditions, then a significantly more complex model must be created, allowing to find a dependence between the indicators of the assessment of the corrosion state from factors ($X_1, X_2, ... X_n$) determining the nature of the development of corrosion processes.

During the research, various functional, informational and morphological models are created, subject to the requirements for reproducibility, and on this basis the overall model of the system is built. Modeling can be complete, incomplete and approximate [98, 99, 103].

Full modeling is carried out by taking into account the ratio (Form.4.3):

$$m = \mu . M_0 \tag{4.3}$$

where: m are the model matrices; μ and Mo are the coefficients and parameters of the original, respectively.

In incomplete modeling, the model is a function similar to the original, generally described by the expression (Form. 4.4):

$$m = \mu . M_0 \;\; ; \;\; M = \rho(M_0 . x, \tau) \tag{4.4}$$

where: ρ is some function of the original, x are spatial coordinates;

In approximate modeling, only the most significant factors of the process are taken into account, in this case corrosion, through a system that generally looks like this (Form. 4.5):

$$\begin{cases} m = \mu . M_0 \\ M = sub\ M_0 \end{cases} \tag{4.5}$$

By their composition, models can be physical, mathematical, with natural elements and mixed.

Physical models are built using the method of similarity and analogy, including interconnected diverse elements. They help to reproduce the corrosion process in laboratory conditions.

Mathematical models reflect the actually occurring corrosion processes with the help of mathematical equations and their graphic representation in the form of tabular information, nomograms, flowcharts, systems with sets of equations connected in a vertical and horizontal hierarchy, forming matrices of solutions for various cases, including cybernetic models, as well as those built on a block principle. This format also includes algorithmic descriptions, which are often used to present a model of objects for which there is no analytical description or the model itself is a preparation for a programmable calculation, as is the case with corrosion processes and is often made directly in a programming language, and these two options may coincide. A variety of mathematical modeling are graphs, especially in their form of target tree structures, which are successfully used in programming [85, 86, 98, 99, 100].

Cybernetic models should be simple and complete. Simple models include static models that do not take into account changes in corrosion factors over time. Complete models take into account the dynamics of changes in corrosion factors over time, as well as the relationship between them. In order to fulfill the main task of cybernetics - system integrity management, it is necessary for a complete model of corrosion processes to include not only the main factors, but also the limitations and relationships between them. To determine the effectiveness of corrosion protection methods, such models must also contain criteria and functions for optimality [98, 99, 100, 103, 106].

Models with full-scale elements have a technical device (block - schemes, means for studying processes and technical complexes in full).

Elements mixed in their composition contain complex means and time synchronizers of the modeling process, which are necessary for studying the use of objects for their intended purpose. In this case, corrosion is one of the factors that reduce the probability that the object will fulfill its re-purpose in time.

In addition, in terms of their description, models of both corrosion and other processes can be deterministic, stochastic, heuristic and mixed.

Deterministic models are set in the form of logical, algebraic and differential equations or their solutions, recorded as a function of time, as well as experimental data obtained in natural conditions or during accelerated corrosion tests.

Probabilistic models are based on operations with random numbers, when the nature of the corrosion process is unknown [78].

Heuristic models are formed on the basis of hypotheses that take into account the possible direction of development of the corrosion process.

Mixed models are a combination of the models already described. Heuristic and mixed models are rarely used, since their expediency in the analysis of corrosion protection options is justified only and only in complex systems with mutually exclusive elements of the popular corrosion protection [78].

Most often, deterministic models are used in the research and diagnostics of corrosion processes. Determining the real value of the corrosion process rate and the change in the rate is part of the subject of the study of the kinetics of corrosion processes and is a leading goal in their modeling. Usually, the corrosion rate [79] is understood as the average rate v_{cp} of the corrosion process (Form. 4.6):

$$V_{cp} = \tau^{-1} \sum_{1}^{n} l_{k,i} \quad (4.6)$$

where: $l_{k,i}$ – the corrosion effect at a given point in time;

The actual rate of corrosion process differs significantly from the average rate and has characteristic features that should not be taken into account in the operating conditions, especially when predicting corrosion hazard.

The corrosion process always starts from the surface of the metal and is characterized by development in depth. The corrosion process itself is the result of various reasons, discussed above, and cannot be uniform [79]. The change in corrosion in depth can be represented in the form of an analytical model by the following formula (Form.4.7):

$$l_k = l_{k,y}\left[1 - exp\left(\frac{\tau}{k_\tau}\right)\right] \quad (4.7)$$

where: $l_{k,y}$ – depth of corrosion at the beginning of the process

$K\tau$ – constant for the special case of corrosion;

The actual rate v_d of the corrosion process is found after differentiating equation [79] (Form. 4.8):

$$v_d = v_0 . \exp\left(\frac{-\tau}{k_\tau}\right) \quad (4.8)$$

The models formed by (1) and (2) describe only a part of the corrosion process – only in the region of the specified values of $l_{k,y}$. To express the actual rate of the corrosion process as a whole from the zero value, the analytical model [72, 79] (Form. 4.9) has been empirically determined:

$$V(\tau) = l'_k(\tau) = v_d . \tau . (k_l . \tau^2 + k_{l'} . \tau + k_{l''})^{-1} \quad (4.9)$$

Where: κ_l , $\kappa_{l'}$, $\kappa_{l''}$ – constant values, inversely proportional to the depth of corrosion penetration into the metal, the speed and acceleration of the corrosion process:

Analysis of model (3) shows that the corrosion rate is a complex nonlinear function, constantly changing in time. At the initial moment, the rate is always equal to zero, which also characterizes the inertia of the corrosion process, but then it increases and slows down. Its maximum value should coincide with the point of interruption of the aperiodic curve, indicating the size of the depth of corrosion, and over time it can be expressed with a certain and established value.

When the calculations are made with the average corrosion rate, which is essentially a fictitious value, the possibility of analyzing the kinetics of the corrosion process is virtually excluded and prevents its diagnosis and prediction.

Recently, in the construction of mathematical models of various physical processes, the identification method has been successfully applied, which represents the construction of the process model based on the primary information obtained under operational conditions [98, 99], which can be represented by the following expression, which determines the approximate value of the linear operator A_0 (Form. 4.10):

$$l_i(\tau) = A_0.x(\tau) \tag{4.10}$$

Particularly important for the definition of the mathematical model is the issue of adaptation. Models that are built on the basis of information about the initial processes can lead to erroneous conclusions about the real corrosion situation. To exclude the possibility of such errors, adaptive models are adopted, allowing correction to be made at a specific point in time in accordance with changing conditions. [98, 99]. When developing a model of corrosion processes, it is important to find the optimal algorithm that provides the necessary qualities of the resulting model, with minimal effort and resources and with maximum use of the accumulated experience. Based on the basis presented in Fig. 4.2, the modeling process is divided into several interconnected stages. In the first stage, the goals and significance of the study are identified using models that evaluate the available information about similar processes and their ability to formalize the processes described by the models, which predetermine the choice of the type and type of model. In essence, the first stage forms the capabilities of the future model to express the process of corrosion, aging and biological damage. In the second stage, the model itself is built, allowing full consideration of the set of factors influencing the process and their complex interaction with each other. If the process can be described with analytical equations, a deterministic model is obtained. If there is no such possibility, then one should move on to a probabilistic description and build a stochastic model [100]. At the third stage, the model is corrected, which is necessary for the following reasons. During the preliminary study of the corrosion process, it may turn out that the capabilities of the mathematical apparatus and computing equipment are insufficient for correct calculation and complete solution of the tasks. There is a need for reasonable simplification of the model, with an assessment of its capabilities. When formalizing individual relationships, either the dependencies used earlier in the analysis of related processes are used, or statistical data obtained when processing information for individual experiments with similar conditions.

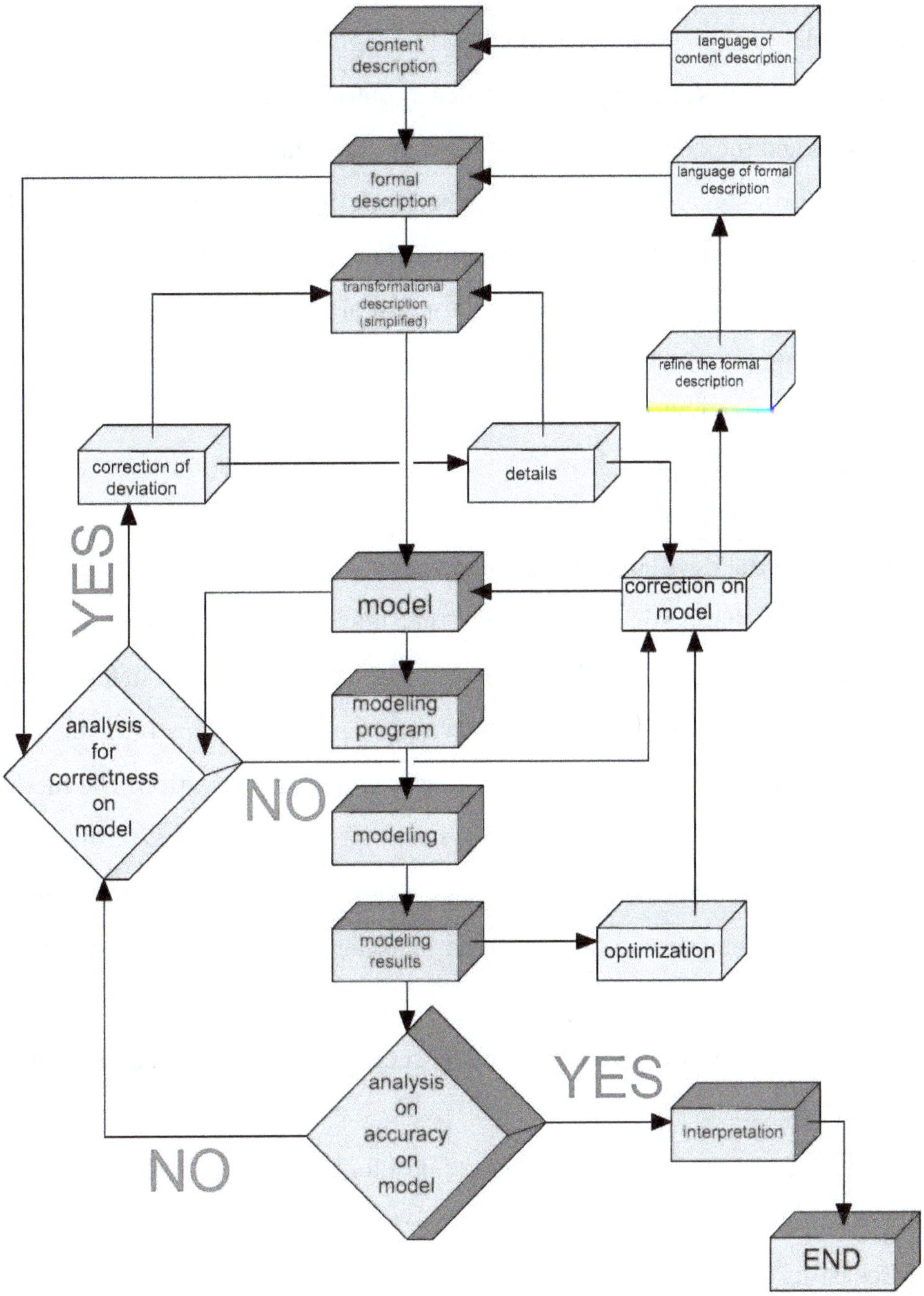

Figure 4.2 Bearing steel elements with corrosion in structures [110]

Naturally, in these cases, the degree of approximation of the results achieved with the help of the models to real values and reliable meanings may be insufficient. For the purposes of corrections, the coincidence of the results is a necessary condition. Finally, in the process of preliminary analysis, weak sensitivity of the model to changes in individual factors significantly affecting the process or excessive sensitivity to secondary factors may appear. In this case, clarification of the formalized description of the constituent processes and the connectivity between individual factors is required.

In the final fourth stage, the accuracy of the modeling as a whole is assessed, specifying the capabilities of the model and developing recommendations for its use. As a rule, in the process of using the model, it is corrected periodically, either automatically, if the model has high adaptability, or according to the algorithm discussed above. Periodic correction is especially important for forecasting models, since the degree to which the model corresponds to the actual process at the beginning of the forecast directly affects the accuracy.

3.4. Forecasting the development of processes

Forecasting is a probabilistic assessment of the state of an object (process or phenomenon) at a certain point in time in the future [98, 99], for example, the state of an object in terms of corrosion, aging and biological damage. In relation to the issue under consideration, forecasting is the process of forming characteristic signs for the development of the processes under consideration. The model obtained from the methods and used for forecasting is called prognostic. It must meet the following requirements:

- to correspond to the intended completeness, adaptability and evolutionism and to allow the introduction of amendments and additions in order to achieve a consistent approach to the required accuracy;
- be abstract enough to allow variation of the required number of factors and to guarantee the reliability of the results obtained;
- to have reasonable deadlines for solving the problem in order to achieve quick results in short-term forecasting and to exclude large costs of machine time for processing the results in long-term forecasting or evaluating the forecast results;
- be oriented towards implementation through well-known technical means and be based on established terminology;
- to ensure the receipt of useful information about the process in relation to the research task;
- to provide opportunities for verification of accuracy in accordance with actual corrosion, aging and biodegradation processes.

The principles and rules that determine the general properties of models are described in the special literature [98, 99].

Depending on the number of variables included in the model description at the analysis stage, the forecasting objects are divided into sublocal (from 1 to 3 factors), local (from 4 to 14 factors), subglobal (from 15 to 35 factors), global (with a total number of significant variables from 35 to 100) and superglobal (with more than 100 factors) [103]. The processes of corrosion, aging and biodegradation as forecasting objects mainly belong to the first three classes. In particular, as mentioned, atmospheric corrosion is a subglobal forecasting object [78]. In addition to the number of variables, the forecasting method is also influenced by the complexity of the object:

- super-simple - these are objects without significant relationships between variables. For them, predictions can be constructed through a sequential analysis of independent variables Xt (factors), which make up the description of objects of any magnitude;
- simple - in the description of which there are dual relationships between the variables, in which cases the model forecasts are created for pairwise regressions, simple expert methods for assessing the degree and nature of the relationships between the variables;
- complex - for the appropriate description of which it is necessary to take into account the interrelation and joint influence of several significant factors - three or more, in which the main influencing factors can be separated. As a mathematical apparatus, methods of multiple regression and correlation analysis, expert tables for assessing mutual influence and preferences are used;
- supercomplex - objects whose description must take into account the relationship between all variables. The main methods for their study are multiple, correlation, factor and variance analyses [103].
- The processes under consideration, depending on their mechanism, the conditions of their occurrence and the degree of research, can be classified in terms of complexity into one of the first three objects.
- By the nature of development over time, forecasting objects are divided into:
- discrete, whose constant component (trend) changes in a jump-like manner at certain moments in time, whereby a constant component is understood as a description of the process that is free from random components or interference;
- aperiodic, having a description of the constant component in the form of an aperiodic continuous function of time.

Corrosion processes are usually related to aperiodic predictive objects and less often to discrete ones. The same can be said for aging and biodegradation.

3.5. Modeling a computing system

The models can be [98, 99]:

- deterministic, described with sufficient accuracy for prediction, whereby random components can be ignored;
- stochastic, in which it is necessary to take into account the random components of the factors, taking into account the necessary accuracy and the requirements of the forecasting task;
- mixed, containing components of a deterministic and stochastic nature.

Deterministic objects assume a priori known laws of the kinetics of corrosion processes. The forecast is carried out on the basis of purely local information, which is received for processing directly from the controlled object.

The obtained parameters are introduced into the existing model through the equations of the kinetics of the process. In linear deterministic systems [98] the principle of superposition is implemented, according to which the effect of the corrosion effect arising from the influence of several factors is equal to the sum of the effects of the influence of each factor separately. The schemes of the main stages of collecting, processing and analyzing information when forecasting deterministic objects and deterministic forecasting of corrosion processes are presented as variants of the situation.

- Option I – by input parameters is shown in Fig. 4.3 and is used to predict the effects of corrosion damage in operational conditions, using parameters characterizing the immediate situation.
- Option II – by input – output parameters is shown in Fig. 4.4. Provides and predicts information about the factors and effects of the specific technical system, as well as the subsequent processing of this information. The controlled parameters are fixed and regulated. The result of the influence of the influencing factors is fixed at the output by external inspection of the structure during maintenance or repair after a certain period of time. The option does not allow for full automation of the system for monitoring the operational situation [78, 79, 80].
- Option III - by initial parameters or simplified is shown in Fig. 4.5. Its implementation is suitable for both trial operation and laboratory conditions. Under real operating conditions this is unacceptable, since the results are associated with the assumption of the development of processes that can lead to failure of the technical system, as well as the occurrence of an emergency situation of the operating equipment.

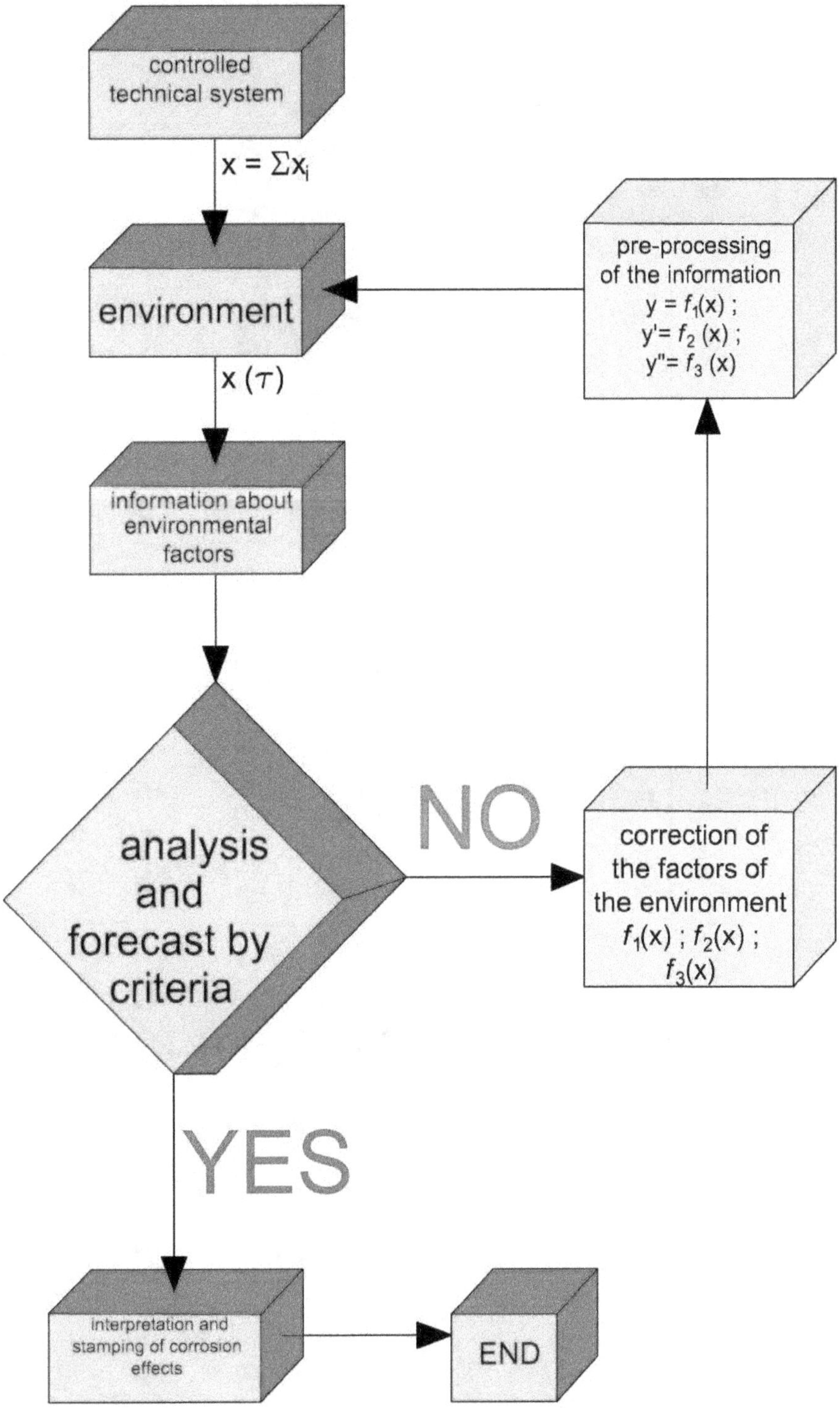

Figure 4.3 Scheme of option I – by input parameters [110]

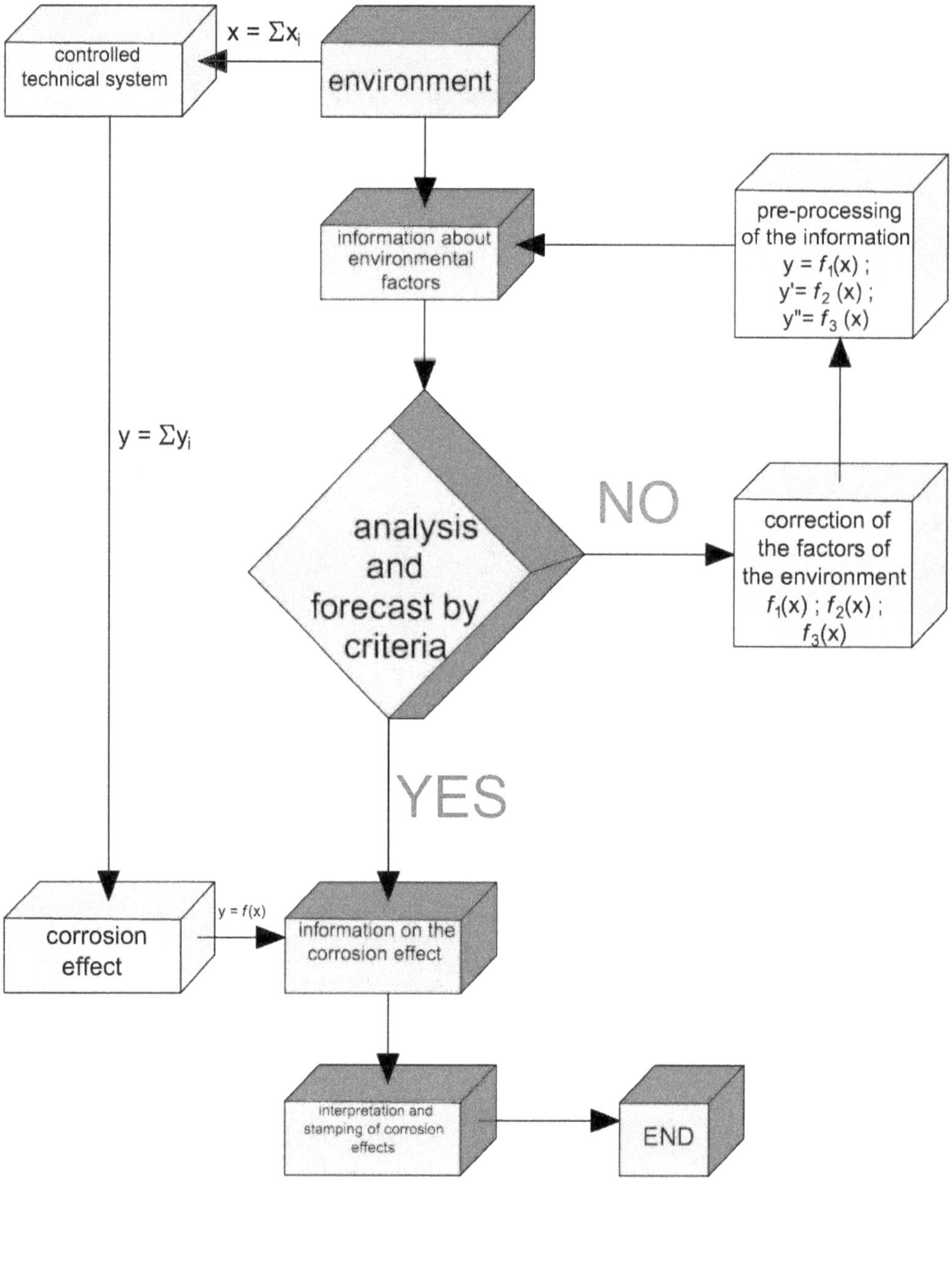

Figure 4.4 Scheme of option II – by input – output parameters [110]

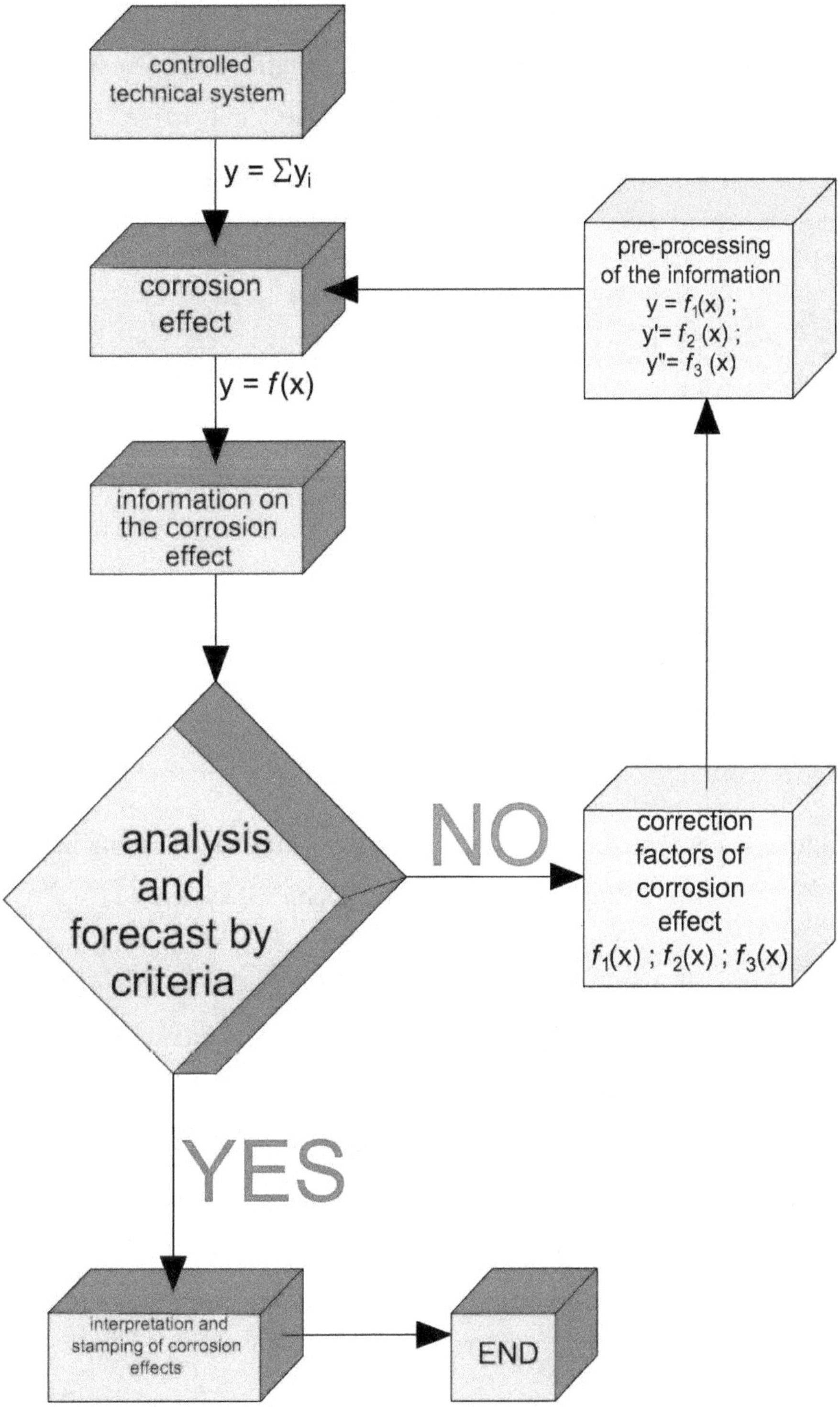

Figure 4.5 Scheme of option III – by initial parameters (simplified) [110]

- Option IV – by initial parameters is presented in Fig. 4.6, taking into account the level of collapse or failure of the technical system. The scheme can be implemented by taking into account the most significant parameters of the external factors affecting corrosion. It is advisable to compare the estimated data with the results of laboratory tests and in natural conditions in order to improve the model of the corrosion process.

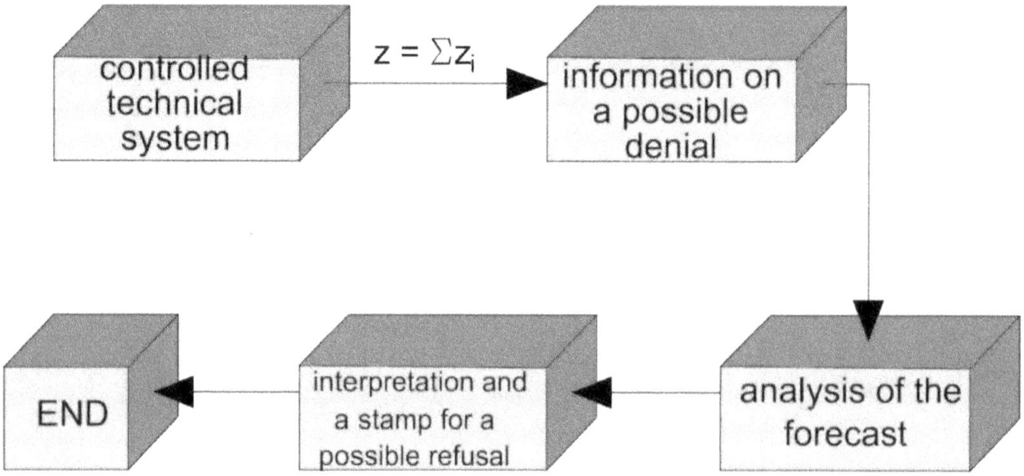

Figure 4.6 Scheme of option IV – by output parameters, taking into account the level of collapse or failure of the technical system [110]

In Fig. 4.7 is a schematic diagram of systems for predicting various types of situations leading to the occurrence of corrosion. Their implementation is much more complicated than the implementation of deterministic objects. The difficulty lies in the fact that it is practically impossible to trace the cause-effect relationship of phenomena that objectively exist in all processes of change in the state of the material, including corrosion processes. Mixed (deterministic-stochastic) objects effectively combine the advantages and eliminate the disadvantages. They are based on a well-studied model of deterministic objects, as well as on the statistical information necessary to determine the specific parameters of the processes.

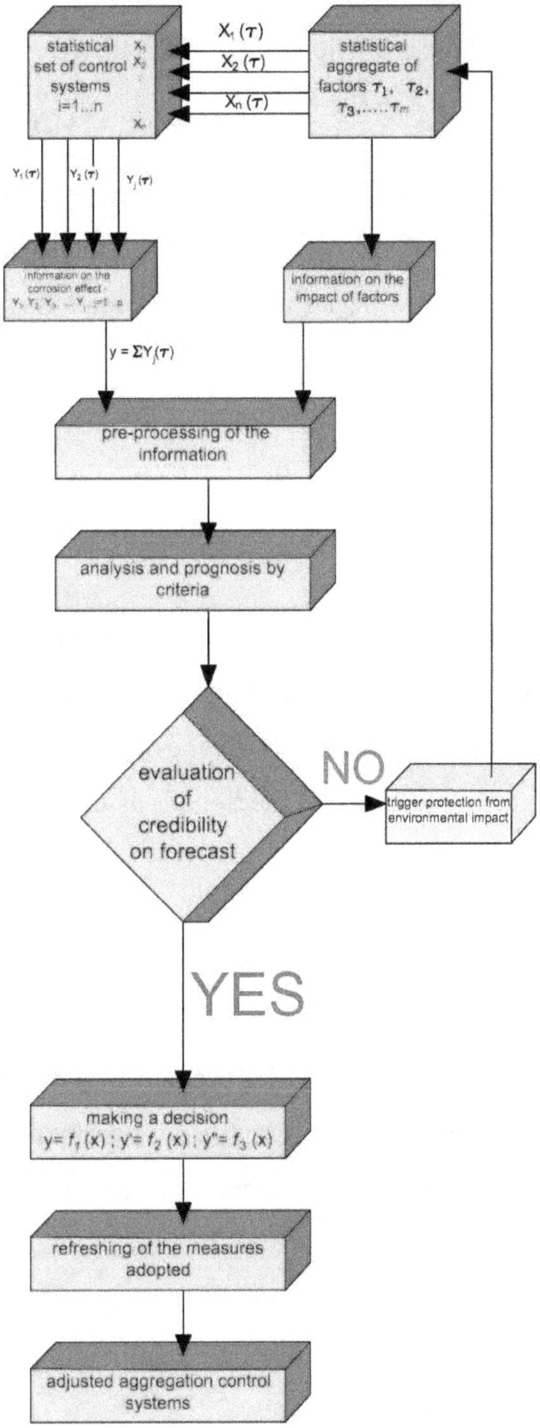

Figure 4.7 Scheme of the main stages of information collection, processing and analysis in stochastic prediction of corrosion processes [110]

3.6. Creating a block-type algorithm

Mixed object models in i(x) are used for extrapolation prediction methods [98, 99, 103] and can be used to model (Fig. 4.11) the corrosion effect – KE(x):

$$KE\ (x) = \ i_n(x) = f\ (a, x) + \ \eta\ (x) \tag{4.11}$$

where: f (a,x) - regular (deterministic) component describing the general trend of change in the damage process over time (trend) and continuing for the forecast period; η (x) is a random (stochastic) component, usually considered to be an uncorrelated random process resulting from unrecorded or insignificant factors with zero mathematical expectation. Its estimates are necessary to determine the accuracy of the forecast [106].

Predictive extrapolations have their own characteristic features and techniques. They include: pre-processing of digital information, including smoothing of empirical data, elimination of random fluctuations and detection of a trend;

transformation of functions to bring them to a point of view convenient for forecasting; analysis of the nature of the predicted process of corrosion, aging and physical damage, revealing its logic and physics, allowing to draw a conclusion about the type of extrapolation and determine the limits of change in its parameters. In this case, the nature of the trend is assessed (increase, decrease, extremes, inflection points, symmetry, restrictions, boundaries).

The sequence of operations for selecting the type of function to describe almost any damage process under the influence of environmental factors and its prediction by extrapolation is given in special literature [103].

The following procedure is adopted for predicting additional polarity:

visual comparative analysis of the graphical representation of the corrosion process (smoothed series) with a limited number of simple extrapolation and interpolation functions using polynomials; automatic selection of the type of extrapolation function;

smoothing of extrapolation and the moving average method. Fig. 4.8 shows a block diagram of the algorithm for calculating predicted values with smoothing of the power polynomial extrapolation used in the description of corrosion processes.

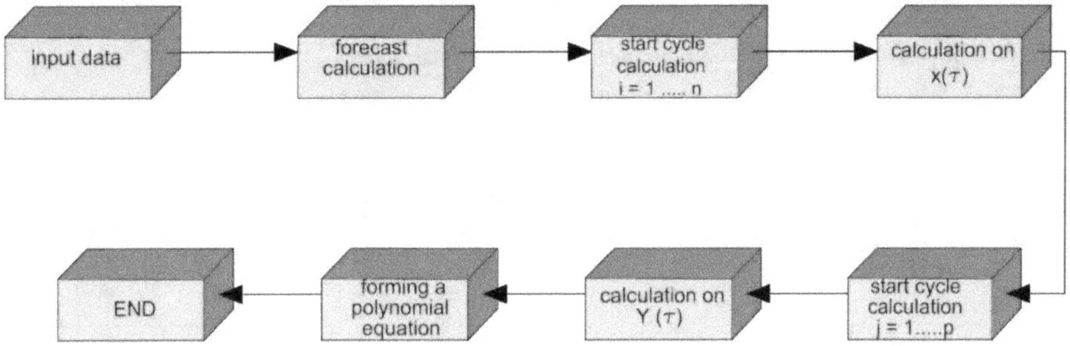

Figure 4.8 Block diagram of the algorithm for calculating the predicted effects of influence due to corrosion [110]

4. Forecasting in sustainable development

The solution of forecasting in sustainable development begins [101, 105]: when the requirements for the material in the product are formulated, the area of its operational usability is determined, when the conditions under which the material will be used in the given product in particular are determined;

- design based on the selected model and corrosion function
- when choosing the material offered in this product;
- when the requirements for reliability and validity of the forecast are formulated.
- The solution to the forecasting problem includes the following step:
- selection of a material aging model according to the parameters included in it;

Determination and refinement of the values of the parameters entering the aging function, using the available information about the material and the product (passive prediction) or special tests (active prediction); formulation of the solution to the prediction problem.

Prediction of sustainable development is the comparative assessment of a material and can be any characteristic of the corrosion function of this type of material. A general approach to this choice is impossible, since it is too diverse, as the corrosion development functions and their influence on stresses, shown in Fig.4.9.

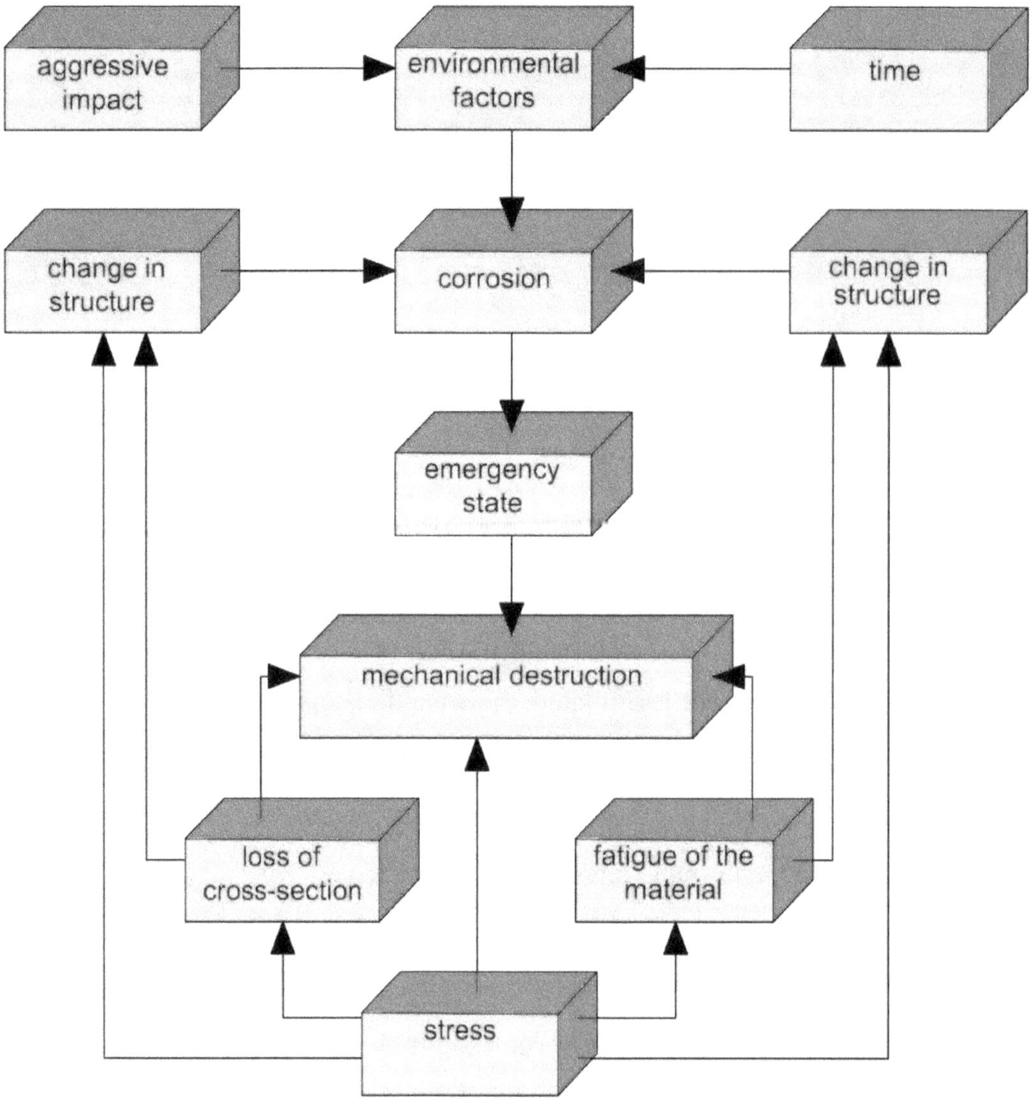

Figure 4.9 Block diagram of the algorithm for calculating the predicted effects of effects due to corrosion [110]

To predict a process that depends on too many aggregate factors is a complex task. The proposed algorithms and models for predicting the development of corrosion, and hence predicting its impact, is an idealistic model based on theory and research. This model should be applied in practical conditions with a view to its refinement and the possible determination of the unknown constants, which will obey the mathematical dependencies reflected in probability theory [103, 104, 106].

It is undeniable that with possible programming of the model, and the participation of a sufficient number of databases for the corresponding development of corrosion in a given metal element, this model will have its practical applicability.

IV.B. Prediction of residual strength and relative deformation

1. Corrosion category C1

The corrosion category is not found in natural climatic conditions, but is intended for indoor climatic conditions in buildings according to the ISO 12944-5 standard, such as heated production buildings, offices, shops, schools, hotels, etc.

1.1 Yield strength

On fig. 4.10(a) shows the relationship between the change in relative deformation versus the time of action of the corrosion category, and Fig. 4.10(b) shows the relationship between the change in strength over time as a result of the corrosion effect of the category.

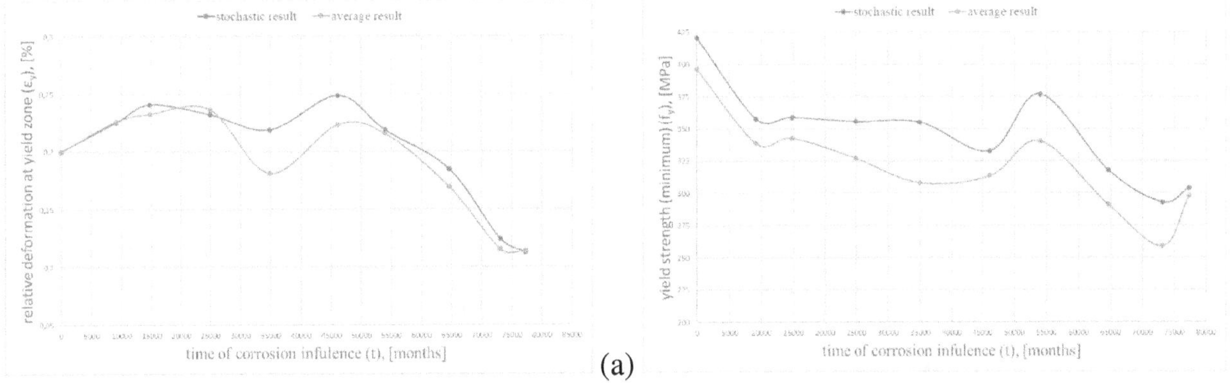

Figure 4.10 (a) graph of the change in relative deformation depending on the corrosion effect over time [108]; (b) graph of the change in strength depending on the corrosion effect over time [109];

From fig. 4.10, using the polynomial approximation [91], a functional dependence is established for the change in strength and relative deformation versus the time of exposure to the corrosion category (in months) depending on the selected data processing method.

a) stochastic result:

• relative deformation at yield zone (eq. 4.12):

$$\varepsilon_y(t) = 1.886923397.10^{-41}.t^9 - 6.607550888.10^{-36}.t^8 + 9.646014499.10^{-31}.t^7 - 7.598353092.10^{-26}.t^6 + 3.492133601.10^{-21}.t^5 - 9.455866912.10^{-17}.t^4 + 1.451318008.10^{-12}.t^3 - 1.148156471.10^{-8}.t^2 + 3.858748796.10^{-5}.t + 2.000583103.10^{-1} \quad (4.12)$$

• yield strength (minimum) (eq. 4.13):

$$f_y(t) = 3.00588407.10^{-38}.t^9 + 1.052798012.10^{-32}.t^8 - 1.54034997.10^{-27}.t^7 + 1.221940811{,}10^{-22}.t^6 - 5.709346323.10^{-18}.t^5 + 1.597993784.10^{-13}.t^4 - 2.603843089.10^{-9}.t^3 + 2.268010889.10^{-5}.t^2 - 8.576219925.10^{-2}.t + 420.8058409 \quad (4.13)$$

b) average result:

- relative deformation at yield zone (eq. 4.14):

$$\varepsilon_y(t) = 2.582118882.10^{-41}.t^9 - 9.4182287.10^{-36}.t^8 + 1.440839012.10^{-30}.t^7 - 1.197855805.10^{-25}.t^6 + 5.858582238.10^{-21}.t^5 - 1.704246908.10^{-16}.t^4 + 2.838036214.10^{-12}.t^3 - 2.447884284.10^{-8}.t^2 + 8.558381421.10^{-5}.t + 2.000560918.10^{-1} \quad (4.14)$$

- yield strength (minimum) (eq. 4.15):

$$f_y(t) = -1.044632037.10^{-38}.t^9 + 3.643583004.10^{-33}.t^8 - 5.309275807.10^{-28}.t^7 + 4.209565499.10^{-23}.t^6 - 1.984743062.10^{-18}.t^5 + 5.711787982.10^{-14}.t^4 - 9.861547308.10^{-10}.t^3 + 9.480843204.10^{-6}.t^2 - 4.275694289.10^{-2}.t + 396.6635496 \quad (4.15)$$

1.2 Tensile strength

On fig. 4.11(a) shows the relationship between the change in relative deformation versus the time of action of the corrosion category, and Fig. 4.11(b) shows the relationship between the change in strength over time as a result of the corrosion effect of the category.

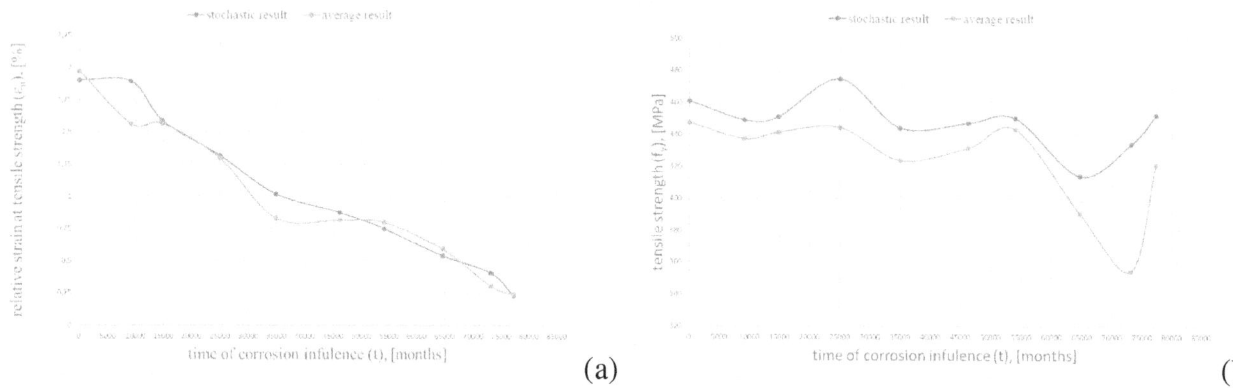

(a) (b)

Figure 4.11 (a) graph of the change in relative deformation depending on the corrosion effect over time [107]; (b) graph of the change in strength depending on the corrosion effect over time [107];

From Fig. 4.11, using the polynomial approximation [91], a functional dependence is established for the change in strength and relative deformation versus the exposure time of the corrosion category (in months) depending on the selected data processing method.

a) stochastic result:

- relative strain at tensile strength (eq. 4.16):

$$\varepsilon_u(t) = 4.300391306.10^{-41}.t^9 - 1.65019893.10^{-35}.t^8 + 2.667269419.10^{-30}.t^7 - 2.359985201.10^{-25}.t^6 + 1.243303922.10^{-20}.t^5 - 3.972351371.10^{-16}.t^4 + 7.491859784.10^{-12}.t^3 - 7.637135628.10^{-8}.t^2 + 3.010721725.10^{-4}.t + 1.902706585 \quad (4.16)$$

- tensile strength (eq. 4.17):

$$f_u(t) = 4.567188873.10^{-39}.t^9 - 1.944151203.10^{-33}.t^8 + 3.422275948.10^{-28}.t^7 - 3.232953691.10^{-23}.t^6 + 1.775685996.10^{-18}.t^5 - 5.726521554.10^{-14}.t^4 + 1.037518148.10^{-9}.t^3 - 9.371899552.10^{-6}.t^2 + 3.08709247.10^{-2}.t + 461.1038654 \quad (4.17)$$

b) average result:

- relative strain at tensile strength (eq. 4.18):

$$\varepsilon_u(t) = 6.161570328.10^{-41}.t^9 - 2.239821925.10^{-35}.t^8 + 3.399410323.10^{-30}.t^7 - 2.779694866.10^{-25}.t^6 + 1.315871686.10^{-20}.t^5 - 3.5941174.10^{-16}.t^4 + 5.29231186.10^{-12}.t^3 - 3.5216038.10^{-8}.t^2 + 3.30542855.10^{-5}.t + 1.973359434 \quad (4.18)$$

- tensile strength (eq. 4.19):

$$f_u(t) = 2.525880003.10^{-40}.t^9 - 1.839988163.10^{-34}.t^8 + 4.541831589.10^{-29}.t^7 - 5.368275473.10^{-24}.t^6 + 3.399263928.10^{-19}.t^5 - 1.178006275.10^{-14}.t^4 + 2.131693936.10^{-10}.t^3 - 1.745600812.10^{-6}.t^2 + 3.980608075.10^{-3}.t + 447.3706767 \quad (4.19)$$

2. Corrosion categoryC2

According to ISO 12944-5, Corrosion category C2 is for environments with low levels of pollution, such as agricultural areas. Unheated buildings where condensation can occur, such as warehouses, depots, sports facilities, etc.

2.1 Yield strength

On fig. 4.12(a) shows the relationship between the change in relative deformation versus the time of action of the corrosion category, and Fig. 4.12(b) shows the relationship between the change in strength over time due to the corrosion effect of the category. From Fig. 4.12, using the polynomial approximation [91], a functional relationship is established for the change in strength and relative deformation versus the time of action of the corrosion category (in months) depending on the selected data processing method.

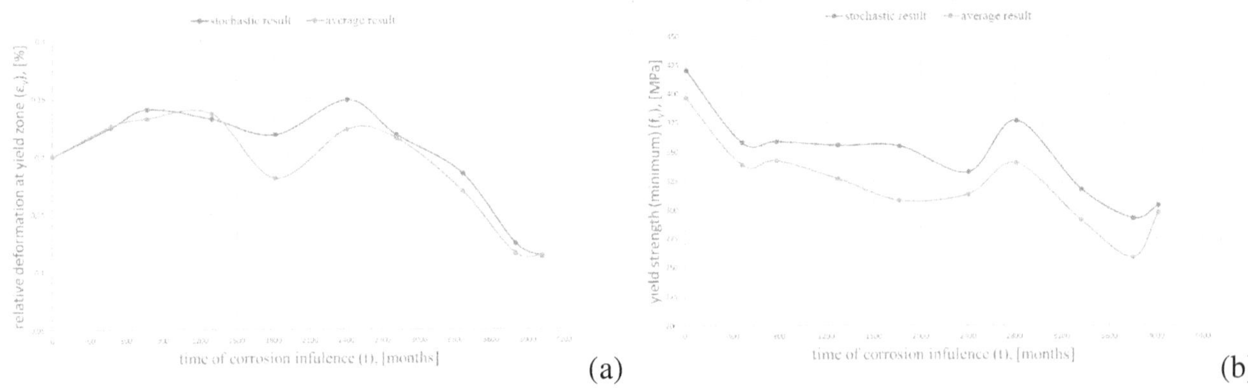

(a) (b)

Figure 4.12 (a) graph of the change in relative deformation depending on the corrosion effect over time [108]; (b) graph of the change in strength depending on the corrosion effect over time [109];

a) stochastic result:

• relative deformation at yield zone (eq. 4.20):

$$\varepsilon_y\ (t) = 6.792692281.10^{-30}.t^9 - 1.23691171.10^{-25}.t^8 + 9.389851881.10^{-22}.t^7 - 3.846326496.10^{-18}.t^6 + 9.192595654.10^{-15}.t^5 - 1.294423135.10^{-11}.t^4 + 1.03317532.10^{-8}.t^3 - 4.250627121.10^{-6}.t^2 + 7.428605068.10^{-4}.t + 1.999931645.10^{-1} \quad (4.20)$$

• yield strength (minimum) (eq. 4.21):

$$f_y\ (t) = -1.082086054.10^{-26}.t^9 + 1.970811614.10^{-22}.t^8 - 1.499448042.10^{-18}.t^7 + 6.185512582.10^{-15}.t^6 - 1.502888569.10^{-11}.t^5 + 2.187413906.10^{-8}.t^4 - 1.853467938.10^{-5}.t^3 + 8.395086108.10^{-3}.t^2 - 1.650622791.t + 420.6569459 \quad (4.21)$$

b) average result:

• relative deformation at yield zone (eq. 4.22):

$$\varepsilon_y\ (t) = 9.29815512.10^{-30}.t^9 - 1.763542935.10^{-25}.t^8 + 1.402906588.10^{-21}.t^7 - 6.064783143.10^{-18}.t^6 + 1.542419991.10^{-14}.t^5 - 2.333163649.10^{-11}.t^4 + 2.020400998.10^{-8}.t^3 - 9.061957531.10^{-6}.t^2 + 1.64756854.10^{-3}.t + 1.999948575.10^{-1} \quad (4.22)$$

• yield strength (minimum) (eq. 4.23):

$$f_y\ (t) = -3.761290085.10^{-27}.t^9 + 6.822088218.10^{-23}.t^8 - 5.169413994.10^{-19}.t^7 + 2.13138301.10^{-15}.t^6 - 5.225701161.10^{-12}.t^5 + 7.820255652.10^{-9}.t^4 - 7.020801584.10^{-6}.t^3 + 3.509508771.10^{-3}.t^2 - 8.226534601.10^{-1}.t + 396.5306315 \quad (4.23)$$

2.2 Tensile strength

On fig. 4.13(a) shows the relationship between the change in relative deformation versus the time of action of the corrosion category, and Fig. 4.13(b) shows the relationship between the change in strength over time due to the corrosion effect of the category. From Fig. 4.13, using the polynomial approximation [91], a functional relationship is established for the change in strength and relative deformation versus the time of action of the corrosion category (in months) depending on the selected data processing method.

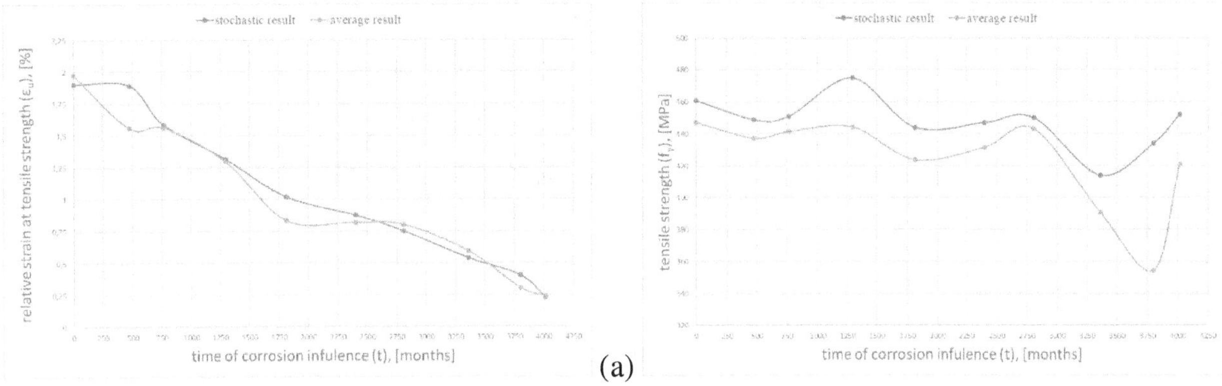

(a) (b)

Figure 4.13 (a) graph of the change in relative deformation depending on the corrosion effect over time [107]; (b) graph of the change in strength depending on the corrosion effect over time [107];

a) stochastic result:

- relative strain at tensile strength (eq. 4.24):

$$\varepsilon_u(t) = 1.548441942.10^{-29}.t^9 - 3.089574242.10^{-25}.t^8 + 2.596598942.10^{-21}.t^7 - 1.19460109.10^{-17}.t^6 + 3.272397952.10^{-14}.t^5 - 5.436387356.10^{-11}.t^4 + 5.33118141.10^{-8}.t^3 - 2.825724466.10^{-5}.t^2 + 5.792219254.10^{-3}.t + 1.902528851 \quad (4.24)$$

- tensile strength (eq. 4.25):

$$f_u(t) = 1.645364274.10^{-27}.t^9 - 3.641210131.10^{-23}.t^8 + 3.332389085.10^{-19}.t^7 - 1.636749255.10^{-15}.t^6 + 4.674167537.10^{-12}.t^5 - 7.837825826.10^{-9}.t^4 + 7.383770237.10^{-6}.t^3 - 3.468278766.10^{-3}.t^2 + 5.943745247.10^{-1}.t + 460.9000627 \quad (4.25)$$

b) average result:

- relative strain at tensile strength (eq. 4.26):

$$\varepsilon_u(t) = 2.218270013.10^{-29}.t^9 - 4.193093337.10^{-25}.t^8 + 3.309209528.10^{-21}.t^7 - 1.407086855.10^{-17}.t^6 + 3.463755929.10^{-14}.t^5 - 4.919823399.10^{-11}.t^4 + 3.767515111.10^{-8}.t^3 - 1.304002088.10^{-5}.t^2 + 6.38555307.10^{-4}.t + 1.973191712 \quad (4.26)$$

- tensile strength (eq. 4.27):

$$f_u(t) = 9.122548483.10^{-29}.t^9 - 3.448603062.10^{-24}.t^8 + 4.423123542.10^{-20}.t^7 - 2.717538314.10^{-16}.t^6 + 8.946267509.10^{-13}.t^5 - 1.611970765.10^{-9}.t^4 + 1.516739114.10^{-6}.t^3 - 6.459404718.10^{-4}.t^2 + 7.687300964.10^{-2}.t + 447.1870387 \quad (4.27)$$

3. Corrosion categoryC3

It exists in urban and industrial regions with moderate sulfur dioxide pollution and in coastal areas with low salinity. In the interior area of a building, it is available in production areas with high humidity and air pollution, such as industrial plants, laundries, breweries, food production areas, etc., according to ISO 12944-5.

3.1 Yield strength

On fig. 4.14(a) shows the relationship between the change in relative deformation and the time of action of the corrosion category, and Fig. 4.14(b) shows the relationship between the change in strength over time due to the corrosion effect of the category.

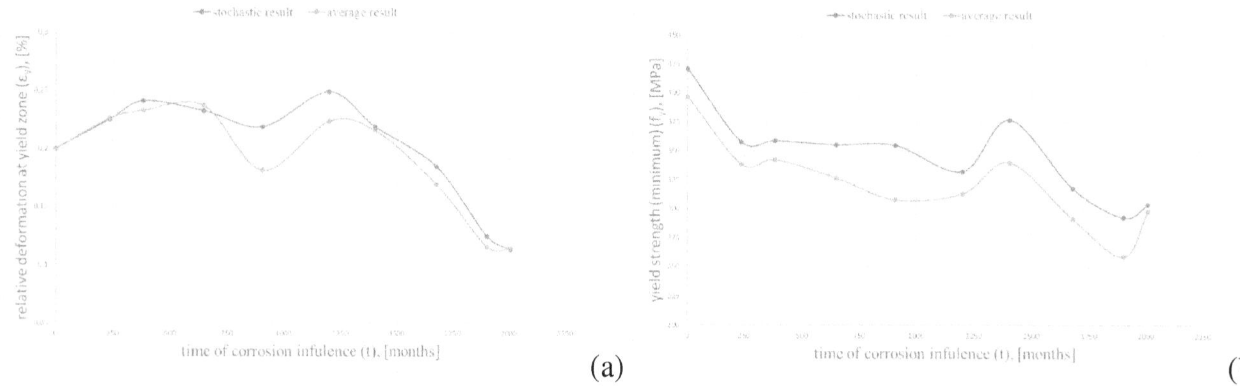

(a) (b)

Figure 4.14 (a) graph of the change in relative deformation depending on the corrosion effect over time [108]; (b) graph of the change in strength depending on the corrosion effect over time [109];

From Fig. 4.14, by using the polynomial approximation [91], a functional dependence is established for the change in strength and relative deformation versus the time of exposure to the corrosion category (in months) depending on the selected data processing method.

a) stochastic result:

• relative deformation at yield zone (eq. 4.28):

$$\varepsilon_y (t) = 3.477858448.10^{-27}.t^9 - 3.166493978.10^{-23}.t^8 + 1.201901041.10^{-19}.t^7 - 2.461648958.10^{-16}.t^6 + 2.941630609.10^{-13}.t^5 - 2.071077015.10^{-10}.t^4 + 8.265402562.10^{-8}.t^3 - 1.700250849.10^{-5}.t^2 + 1.485721014.10^{-3}.t + 1.999931645 \cdot 10^{-1} \quad (4.28)$$

- yield strength (minimum) (eq. 4.29):

$$f_y (t) = -5.540280598.10^{-24}.t^9 + 5.045277731.10^{-20}.t^8 - 1.919293494.10^{-16}.t^7 + 3.958728053.10^{-13}.t^6 - 4.80924342.10^{-10}.t^5 + 3.49986225.10^{-7}.t^4 - 1.482774351.10^{-4}.t^3 + 3.358034443.10^{-2}.t^2 - 3.301245582.t + 420.6569459 \quad (4.29)$$

b) average result:

- relative deformation at yield zone (eq. 4.30):

$$\varepsilon_y (t) = 4.755886101.10^{-27}.t^9 - 4.510221012.10^{-23}.t^8 + 1.793979446.10^{-19}.t^7 - 3.877751853.10^{-16}.t^6 + 4.931074314.10^{-13}.t^5 - 3.729538961.10^{-10}.t^4 + 1.614782559.10^{-7}.t^3 - 3.621263097.10^{-5}.t^2 + 3.291907648.10^{-3}.t + 1.999816201 \cdot 10^{-1} \quad (4.30)$$

- yield strength (minimum) (eq. 4.31):

$$f_y (t) = -1.9240226.10^{-24}.t^9 + 1.744814773.10^{-20}.t^8 - 6.610432844.10^{-17}.t^7 + 1.362717923.10^{-13}.t^6 - 1.670503029.10^{-10}.t^5 + 1.249941921.10^{-7}.t^4 - 5.610989135.10^{-5}.t^3 + 1.402556155.10^{-2}.t^2 - 1.644354763.t + 396.5035214 \quad (4.31)$$

3.2 Tensile strength

On fig. 4.15(a) shows the relationship between the change in relative deformation versus the time of action of the corrosion category, and Fig. 4.15(b) shows the relationship between the change in strength over time due to the corrosion effect of the category.

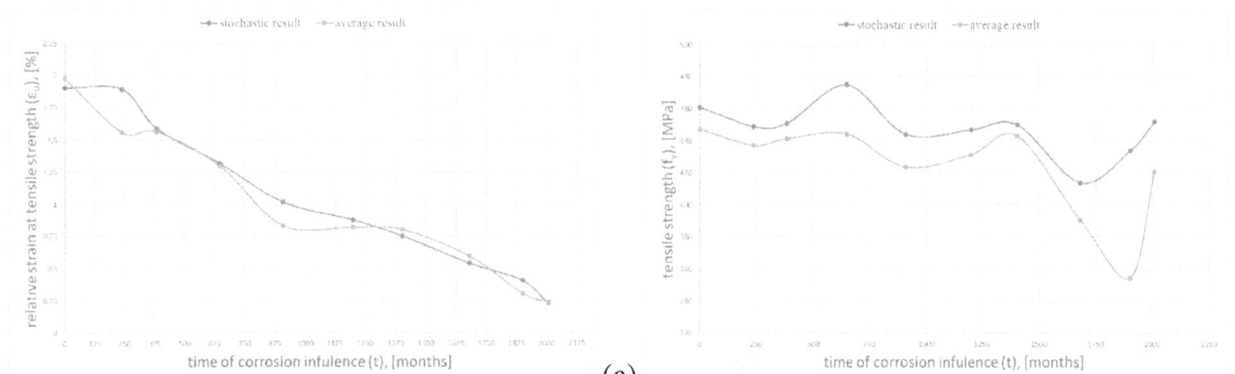

(a) (b)

Figure 4.15 (a) graph of the change in relative deformation depending on the corrosion effect over time [107]; (b) graph of the change in strength depending on the corrosion effect over time [107];

From fig. 4.15, using the polynomial approximation [91], a functional relationship is established for the change in strength and relative deformation versus the time of action of the corrosion category (in months) depending on the selected data processing method.

a) stochastic result:

- relative strain at tensile strength (eq. 4.32):

$$\varepsilon_u\,(t) = 7.928022742.10^{-27}.t^9 - 7.909310059.10^{-23}.t^8 + 3.323646646.10^{-19}.t^7 - 7.645446979.10^{-16}.t^6 + 1.047167345.10^{-12}.t^5 - 8.69821977.10^{-10}.t^4 + 4.264945128.10^{-7}.t^3 - 1.130289786.10^{-4}.t^2 + 1.158443851.10^{-2}.t + 1.902528851 \quad (4.32)$$

- tensile strength (eq. 4.33):

$$f_u\,(t) = 8.412784552.10^{-25}.t^9 - 9.310787721.10^{-21}.t^8 + 4.261266584.10^{-17}.t^7 - 1.046626366.10^{-13}.t^6 + 1.494609337.10^{-10}.t^5 - 1.253204301.10^{-7}.t^4 + 5.903288809.10^{-5}.t^3 - 1.386401322.10^{-2}.t^2 + 1.187672313.t + 460.8595093 \quad (4.33)$$

b) average result:

- relative strain at tensile strength (eq. 4.34):

$$\varepsilon_u\,(t) = 1.135754247.10^{-26}.t^9 - 1.073431894.10^{-22}.t^8 + 4.235788196.10^{-19}.t^7 - 9.005355869.10^{-16}.t^6 + 1.108401897.10^{-12}.t^5 - 7.871717438.10^{-10}.t^4 + 3.014012089.10^{-7}.t^3 - 5.216008353.10^{-5}.t^2 + 1.277110614.10^{-3}.t + 1.973191712 \quad (4.34)$$

- tensile strength (eq. 4.35):

$$f_u\,(t) = 4.668501948.10^{-26}.t^9 - 8.826285475.10^{-22}.t^8 + 5.660739384.10^{-18}.t^7 - 1.73903498.10^{-14}.t^6 + 2.862555787.10^{-11}.t^5 - 2.578950387.10^{-8}.t^4 + 1.213259191.10^{-5}.t^3 - 0.002582563.t^2 + 0.153086761.t + 447.1776509 \quad (4.35)$$

4. Corrosion categoryC4

The ISO 12944-5 standard has defined the existence of this category, in coastal climate and industrial areas with moderate salinity, chemical industry, shipbuilding and vessels.

.

4.1 Yield strength

On fig. 4.16(a) shows the relationship between the change in relative deformation versus the time of action of the corrosion category, and Fig. 4.16(b) shows the relationship between the change in strength over time due to the corrosion effect of the category. From Fig. 4.16, using the polynomial approximation [91], a functional relationship is established for the change in strength and relative deformation versus the time of action of the corrosion category (in months) depending on the selected data processing method.

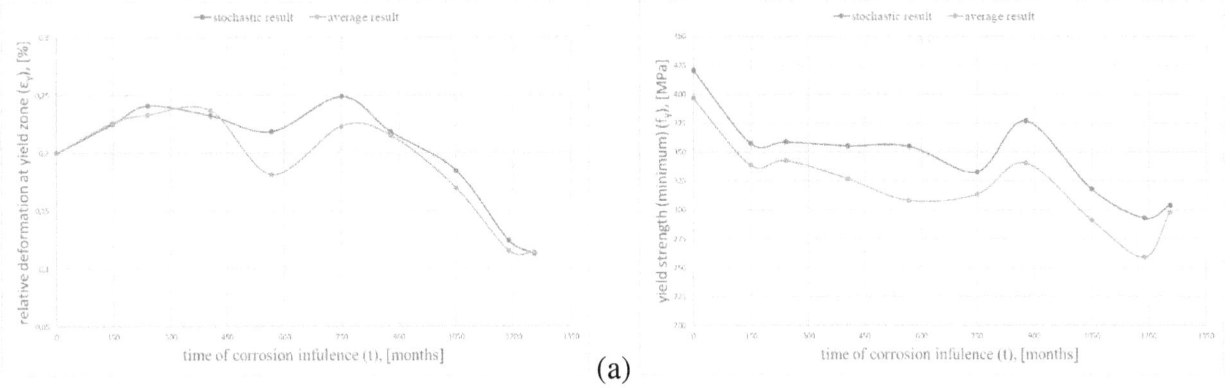

(a) (b)

Figure 4.16 (a) graph of the change in relative deformation depending on the corrosion effect over time [108]; (b) graph of the change in strength depending on the corrosion effect over time [109];

a) stochastic result:

• relative deformation at yield zone (eq. 4.36):

$$\varepsilon_y\,(t) = 2.390889918.10^{-25}.t^9 - 1.36053936.10^{-21}.t^8 + 3.22765619.10^{-18}.t^7 - 4.131738197.10^{-15}.t^6 + 3.08592909.10^{-12}.t^5 - 1.3579682.10^{-9}.t^4 + 3.38734127.10^{-7}.t^3 - 4.355299626.10^{-5}.t^2 + 2.37890968.10^{-3}\,t + 1.999707759\cdot 10^{-1} \quad (4.36)$$

• yield strength (minimum) (eq. 4.37):

$$f_y\,(t) = -3.808744784.10^{-22}.t^9 + 2.167803856.10^{-18}.t^8 - 5.154209694.10^{-15}.t^7 + 6.644513792.10^{-12}.t^6 - 5.045127362.10^{-9}.t^5 + 2.294746695.10^{-6}.t^4 - 6.076402553.10^{-4}.t^3 + 8.600745241.10^{-2}.t^2 - 5.283785272.t + 420.6072564 \quad (4.37)$$

b) average result:

• relative deformation at yield zone (eq. 4.38):

$$\varepsilon_y\,(t) = 3.267501233.10^{-25}.t^9 - 1.93670531.10^{-21}.t^8 + 4.814647795.10^{-18}.t^7 - 6.504414361.10^{-15}.t^6 + 5.169522411.10^{-12}.t^5 - 2.443677952.10^{-9}.t^4 + 6.612742.10^{-7}.t^3 - 9.2683586.10^{-5}.t^2 + 5.26549618.10^{-3}\,t + 2.000378086\cdot 10^{-1} \quad (4.38)$$

• yield strength (minimum) (eq. 4.39):

$$f_y\,(t) = -1.323639783.10^{-22}.t^9 + 7.502438126.10^{-19}.t^8 - 1.776555862.10^{-15}.t^7 + 2.289030365.10^{-12}.t^6 - 1.75382593.10^{-9}.t^5 + 8.201874732.10^{-7}.t^4 - 2.301060012.10^{-4}.t^3 + 3.594522377.10^{-2}.t^2 - 2.63303505.t + 396.4775378 \quad (4.39)$$

4.2 Tensile strength

On fig. 4.17(a) shows the relationship between the change in relative deformation versus the time of action of the corrosion category, and Fig. 4.17(b) shows the relationship between the change in strength over time due to the corrosion effect of the category. From Fig. 4.17, using the polynomial approximation [91], a functional relationship is established for the change in strength and relative deformation versus the time of action of the corrosion category (in months) depending on the selected data processing method.

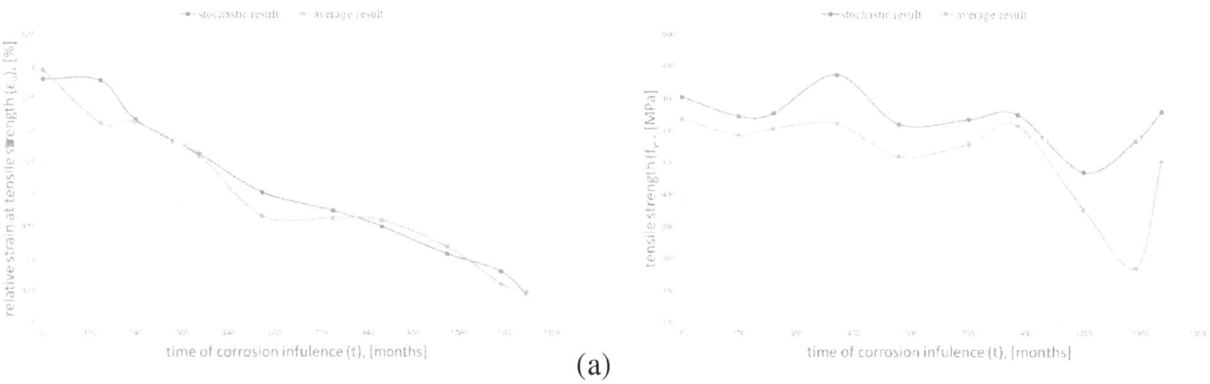

(a) (b)

Figure 4.17 (a) graph of the change in relative deformation depending on the corrosion effect over time [107]; (b) graph of the change in strength depending on the corrosion effect over time [107];

a) stochastic result:

• relative strain at tensile strength (eq. 4.40):

$$\varepsilon_u\ (t) = 5.45086278.10^{-25}.t^9 - 3.398641753.10^{-21}.t^8 + 8.925821873.10^{-18}.t^7 - 1.283225109.10^{-14}.t^6 + 1.098455418.10^{-11}.t^5 - 5.702462184.10^{-9}.t^4 + 1.747469893.10^{-6}.t^3 - 2.894336336.10^{-4}.t^2 + 1.854029757.10^{-2}.t + 1.902466422 \quad (4.40)$$

• tensile strength (eq. 4.41):

$$f_u\ (t) = 5.779625236.10^{-23}.t^9 - 3.998021118.10^{-19}.t^8 + 1.143646822.10^{-15}.t^7 - 1.755643381.10^{-12}.t^6 + 1.566971716.10^{-9}.t^5 - 8.211864007.10^{-7}.t^4 + 2.41766385.10^{-4}.t^3 - 3.548515087.10^{-2}.t^2 + 1.898626957.t + 461.0399536 \quad (4.41)$$

b) average result:

• relative strain at tensile strength (eq. 4.42):

$$\varepsilon_u\,(t) = 7.808116126.10^{-25}.t^9 - 4.612269795.10^{-21}.t^8 + 1.137506092.10^{-17}.t^7 - 1.511477448.10^{-14}.t^6 + 1.162741735.10^{-11}.t^5 - 5.161172686.10^{-9}.t^4 + 1.235188752.10^{-6}.t^3 - 1.336237386.10^{-4}.t^2 + 2.049356457.10^{-3}.t + 1.973132437 \quad (4.42)$$

• tensile strength (eq. 4.43):

$$f_u\,(t) = 3.197066486.10^{-24}.t^9 - 3.784406071.10^{-20}.t^8 + 1.517969995.10^{-16}.t^7 - 2.915533043.10^{-13}.t^6 + 2.999983352.10^{-10}.t^5 - 1.689384934.10^{-7}.t^4 + 4.967411633.10^{-5}.t^3 - 6.607358751.10^{-3}.t^2 + 2.440773845.10^{-1}.t + 447.3132612 \quad (4.43)$$

5. Corrosion category C5

Corrosion category C5 is defined by the ISO 12944-5 standard to be applied to industrial areas with high humidity and aggressive environments. Buildings and areas with constant condensation and pollution above 50% of the pollution in cities.

5.1 Yield strength

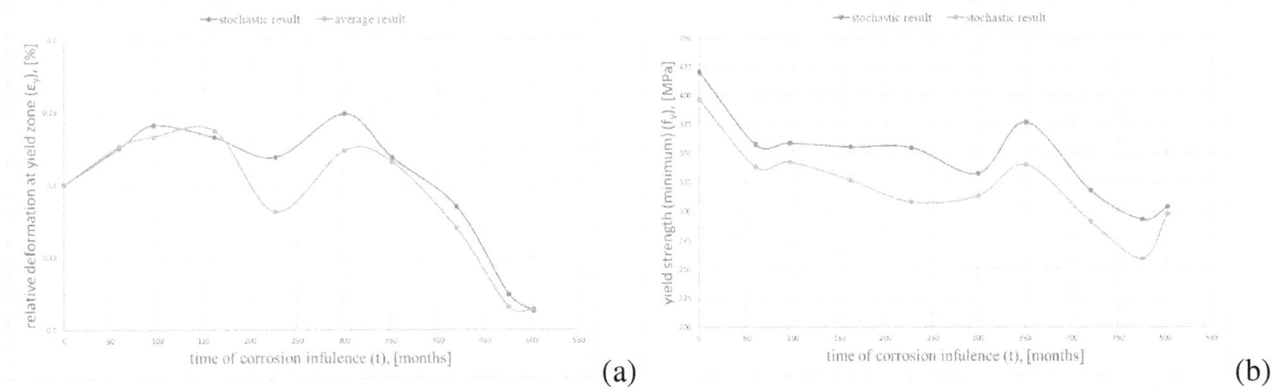

(a) (b)

Figure 4.18 (a) graph of the change in relative deformation depending on the corrosion effect over time [108]; (b) graph of the change in strength depending on the corrosion effect over time [109];

On fig. 4.18(a) shows the relationship between the change in relative deformation versus the time of action of the corrosion category, and Fig. 4.18(b) shows the relationship between the change in strength over time due to the corrosion effect of the category. From Fig. 4.18, using the polynomial approximation [91], a functional relationship is established for the change in strength and relative deformation versus the time of action of the corrosion category (in months) depending on the selected data processing method.

a) stochastic result:

• relative deformation at yield zone (eq. 4.44):

$$\varepsilon_y (t) = 9.116997249.10^{-22}.t^9 - 2.075193493.10^{-18}.t^8 + 1.969194665.10^{-15}.t^7 - 1.008291413.10^{-12}.t^6 + 3.0122297.10^{-10}.t^5 - 5.301957159.10^{-8}.t^4 + 5.2898576.10^{-6}.t^3 - 2.720401358.10^{-4}.t^2 + 5.94288405.10^{-3}.t + 1.999931645 \cdot 10^{-1} \quad (4.44)$$

• yield strength (minimum) (eq. 4.45):

$$f_y (t) = -1.452351317.10^{-18}.t^9 + 3.306473214.10^{-15}.t^8 - 3.14457046.10^{-12}.t^7 + 1.62149501.10^{-9}.t^6 - 4.924665262.10^{-7}.t^5 + 8.95964736.10^{-5}.t^4 - 9.489755845.10^{-3}.t^3 + 5.372855109.10^{-1}.t^2 - 13.20498233.t + 420.6569459 \quad (4.45)$$

b) average result:

• relative deformation at yield zone (eq. 4.46):

$$\varepsilon_y (t) = 1.247717308.10^{-21}.t^9 - 2.958120994.10^{-18}.t^8 + 2.941500762.10^{-15}.t^7 - 1.589518018.10^{-12}.t^6 + 5.053150087.10^{-10}.t^5 - 9.554615085.10^{-8}.t^4 + 1.034218832.10^{-5}.t^3 - 5.798285.10^{-4}.t^2 + 1.3176669.10^{-2}.t + 1.999933406 \cdot 10^{-1} \quad (4.46)$$

• yield strength (minimum) (eq. 4.47):

$$f_y (t) = -5.047302989.10^{-19}.t^9 + 1.144317312.10^{-15}.t^8 - 1.083867953.10^{-12}.t^7 + 5.58601514.10^{-10}.t^6 - 1.711949773.10^{-7}.t^5 + 3.202394851.10^{-5}.t^4 - 3.593798658.10^{-3}.t^3 + 2.245692995.10^{-1}.t^2 - 6.581918497.t + 396.5240525 \quad (4.47)$$

5.2 Tensile strength

On fig. 4.19(a) shows the relationship between the change in the relative deformation versus the time of action of the corrosion category, and Fig. 4.19(b) shows the relationship between the change in strength over time due to the corrosion effect of the category. From Fig. 4.19, using the polynomial approximation [91], a functional relationship is established for the change in strength and relative deformation versus the time of action of the corrosion category (in months) depending on the selected data processing method.

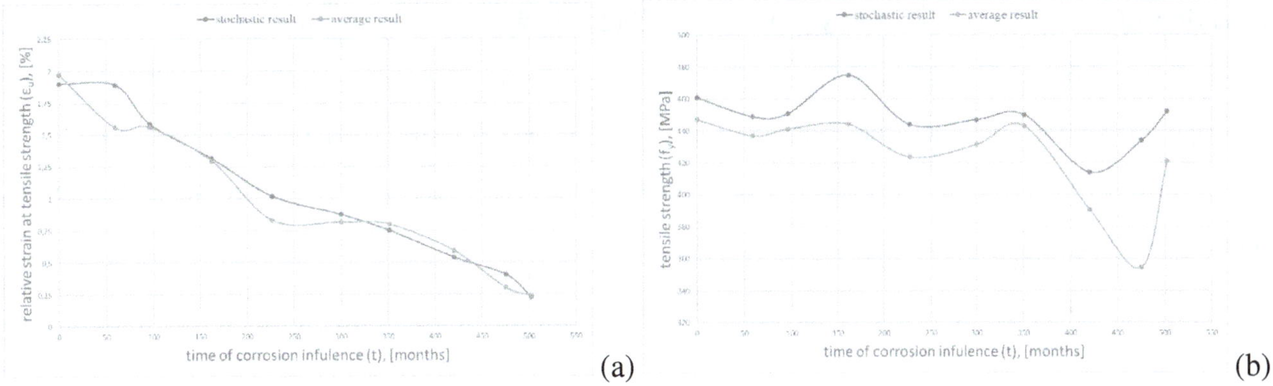

Figure 4.19 (a) graph of the change in relative deformation depending on the corrosion effect over time [107]; (b) graph of the change in strength depending on the corrosion effect over time [107];

a) stochastic result:

• relative strain at tensile strength (eq. 4.48):

$$\varepsilon_u\ (t) = 2.078283594.10^{-21}.t^{9} - 5.18344544.10^{-18}.t^{8} + 5.445462665.10^{-15}.t^{7} - 3.131575082.10^{-12}.t^{6} + 1.072299361.10^{-9}.t^{5} - 2.226744261.10^{-7}.t^{4} + 2.729564882.10^{-5}.t^{3} - 1.808463658.10^{-3}.t^{2} + 4.633775403.10^{-2}.t + 1.902528851 \quad (4.48)$$

• tensile strength (eq. 4.49):

$$f_u\ (t) = 2.204714117.10^{-19}.t^{9} - 6.100408791.10^{-16}.t^{8} + 6.980182659.10^{-13}.t^{7} - 4.286195238.10^{-10}.t^{6} + 1.530232374.10^{-7}.t^{5} - 3.207733895.10^{-5}.t^{4} + 3.777577986.10^{-3}.t^{3} - 2.217892484.10^{-1}.t^{2} + 4.749452975.t + 460.824883 \quad (4.49)$$

b) average result:

• relative strain at tensile strength (eq. 4.50):

$$\varepsilon_u\ (t) = 2.974636691.10^{-21}.t^{9} - 7.028621433.10^{-18}.t^{8} + 6.933846836.10^{-15}.t^{7} - 3.685372869.10^{-12}.t^{6} + 1.133994.10^{-9}.t^{5} - 2.01326509.10^{-7}.t^{4} + 1.926911876.10^{-5}.t^{3} - 8.333989589.10^{-4}.t^{2} + 5.08325535.10^{-3}.t + 1.973141129 \quad (4.50)$$

• tensile strength (eq. 4.51):

$$f_u\ (t) = 1.223819775.10^{-20}.t^{9} - 5.784394449.10^{-17}.t^{8} + 9.274555408.10^{-14}.t^{7} - 7.123087276.10^{-11}.t^{6} + 2.931257126.10^{-8}.t^{5} - 6.60211299.10^{-6}.t^{4} + 7.764858819.10^{-4}.t^{3} - 0.041321008.t^{2} + 6.123470441.10^{-1}.t + 447.1776509 \quad (4.51)$$

6. Corrosion category CX

The ISO 12944-5 defines this category as a coastal area with offshore installations and high salt content. On fig. 4.20(a) shows the relationship between the change in relative deformation and the time of action of the corrosion category, and Fig. 4.20(b) shows the relationship between the change in strength over time due to the corrosion effect of the category. From Fig. 4.20, using the polynomial approximation [91], a functional relationship is established for the change in strength and relative deformation versus the time of action of the corrosion category (in months) depending on the selected data processing method [107].

6.1 Yield strength

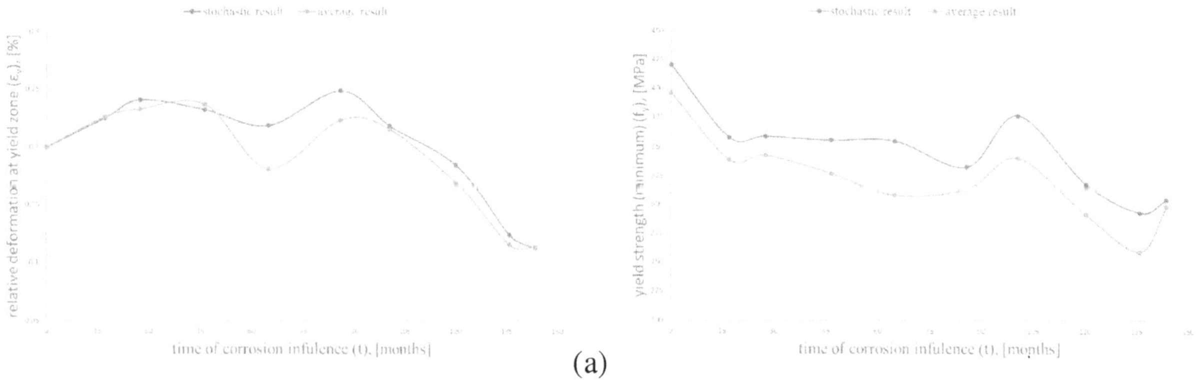

(a) (b)

Figure 4.20 (a) graph of the change in relative deformation depending on the corrosion effect over time [108]; (b) graph of the change in strength depending on the corrosion effect over time [109];

a) stochastic result:

• relative deformation at yield zone (eq. 4.52):

$$\varepsilon_y\ (t) = 7.180920039.10^{-17}.t^9 - 4.669946176.10^{-14}.t^8 + 1.266090957.10^{-11}.t^7 - 1.852176099.10^{-9}.t^6 + 1.5808829.10^{-7}.t^5 - 7.949805788.10^{-6}.t^4 + 2.266021967.10^{-4}.t^3 - 3.329275656.10^{-3}.t^2 + 2.0781186.10^{-2}.t + 1.999960493.10^{-1} \quad (4.52)$$

• yield strength (minimum) (eq. 4.53):

$$f_y\ (t) = -1.14392845.10^{-13}.t^9 + 7.440758881.10^{-11}.t^8 - 2.021795185.10^{-8}.t^7 + 2.978609134.10^{-6}.t^6 - 2.58461175.10^{-4}.t^5 + 1.343475037.10^{-2}.t^4 - 4.065509846.10^{-1}.t^3 + 6.576414205.t^2 - 46.17949256.t + 420.6637067 \quad (4.53)$$

b) average result:

• relative deformation at yield zone (eq. 4.54):

$$\varepsilon_y\ (t) = 9.815281367.10^{-17}.t^9 - 6.648875153.10^{-14}.t^8 + 1.889063452.10^{-11}.t^7 - 2.91666631.10^{-9}.t^6 + 2.649254898.10^{-7}.t^5 - 1.43122937.10^{-5}.t^4 + 4.42624359.10^{-4}.t^3 - 7.089989697.10^{-3}.t^2 + 4.603765.10^{-2}.t + 1.999593074.10^{-1} \quad (4.54)$$

• yield strength (minimum) (eq. 4.55):

$$f_y\ (t) = -3.971483627.10^{-14}.t^9 + 2.572476626.10^{-11}.t^8 - 6.961281746.10^{-9}.t^7 + 1.024994591.10^{-6}.t^6 - 8.974729711.10^{-5}.t^5 + 4.796564155.10^{-3}.t^4 - 1.538018502.10^{-1}.t^3 + 2.746300702.t^2 - 23.00253745.t + 396.4510612 \quad (4.55)$$

6.2 Tensile strength

On fig. 4.21(a) shows the relationship between the change in relative deformation versus the time of action of the corrosion category, and fig. 4.21(b) shows the relationship between the change in strength over time due to the corrosion effect of the category. From fig. 4.21, using the polynomial approximation [91], a functional relationship is established for the change in strength and relative deformation versus the time of action of the corrosion category (in months) depending on the selected data processing method [107].

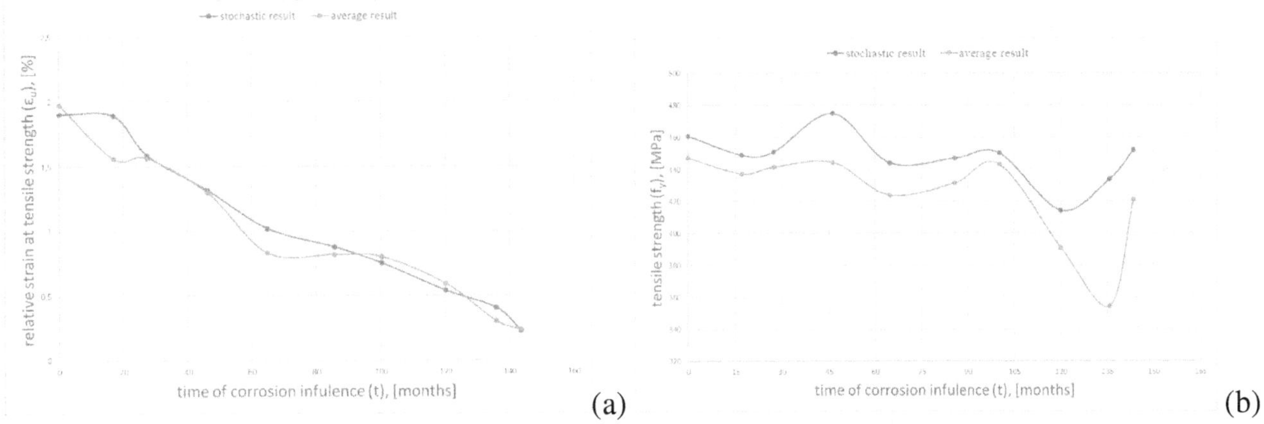

Figure 4.21 (a) graph of the change in relative deformation depending on the corrosion effect over time [107]; (b) graph of the change in strength depending on the corrosion effect over time [107];

a) stochastic result:

• relative strain at tensile strength (eq. 4.56):

$$\varepsilon_u\ (t) = 1.636582059.10^{-16}.t^9 - 1.166294349.10^{-13}.t^8 + 3.50090905.10^{-11}.t^7 - 5.752614746.10^{-9}.t^6 + 5.628275319.10^{-7}.t^5 - 3.339546908.10^{-5}.t^4 + 1.169690779.10^{-3}.t^3 - 2.214387894.10^{-2}.t^2 + 1.621240644.10^{-1}.t + 1.902536082 \quad (4.56)$$

• tensile strength (eq. 4.57):

$$f_u\ (t) = 1.738115986.10^{-14}.t^9 - 1.374036524.10^{-11}.t^8 + 4.491828381.10^{-9}.t^7 - 7.880405146.10^{-7}.t^6 + 8.038163082.10^{-5}.t^5 - 4.814188134.10^{-3}.t^4 + 1.619825107.10^{-1}.t^3 - 2.7173393.t^2 + 16.62781185.t + 460.8995351 \quad (4.57)$$

b) average result:

• relative strain at tensile strength (eq. 4.58):

ε_u (t) = $2.344858317.10^{-16}.t^9$ - $1.583007264.10^{-13}.t^8$ + $4.461873874.10^{-11}.t^7$ - $6.775717808.10^{-9}.t^6$ + $5.956849655.10^{-7}.t^5$ - $3.0216285.10^{-5}.t^4$ + $8.263072059.10^{-4}.t^3$ - $1.02114967.10^{-2}.t^2$ + $1.780792068.10^{-2}.t$ + 1.973198448 (4.58)

• tensile strength (eq. 4.58):

f_u (t) = $9.612900156.10^{-16}.t^9$ - $1.30039439.10^{-12}.t^8$ + $5.961097449.10^{-10}.t^7$ - $1.308489086.10^{-7}.t^6$ + $1.538727554.10^{-5}.t^5$ - $9.902929742.10^{-4}.t^4$ + $3.327917168.10^{-2}.t^3$ - $5.060988776.10^{-1}.t^2$ + 2.147562057.t + 447.1868122 (4.59)

7. Analysis of results

From the established equations (eq. 4.12 – eq. 4.59), it is found that the dependence of the influence of corrosion on the mechanical properties of S355JR steel is a function of a 9th degree polynomial, as follows:

a) Yield strength

• relative deformation at yield zone:

$$\varepsilon_{y,c}(t) = A_1.t^9 + A_2.t^8 + A_3.t^7 + A_4.t^6 + A_5.t^5 + A_6.t^4 + A_7.t^3 + A_8.t^2 + A_9.t + \varepsilon_y$$

Where: A_1, A_2, A_3, A_4, A_5, A_6, A_7, A_8 и A_9 are coefficients.

• yield strength (minimum):

$$f_{y,c}(t) = B_1.t^9 + B_2.t^8 + B_3.t^7 + B_4.t^6 + B_5.t^5 + B_6.t^4 + B_7.t^3 + B_8.t^2 + B_9.t + f_y$$

Where: B_1, B_2, B_3, B_4, B_5, B_6, B_7, B_8 и B_9 are coefficients.

b) tensile strength

• relative strain at tensile strength:

$$\varepsilon_{u,c}(t) = C_1.t^9 + C_2.t^8 + C_3.t^7 + C_4.t^6 + C_5.t^5 + C_6.t^4 + C_7.t^3 + C_8.t^2 + C_9.t + \varepsilon_u$$

Where: C_1, C_2, C_3, C_4, C_5, C_6, C_7, C_8 и C_9 are coefficients.

• tensile strength:

$$f_{u,c}(t) = D_1.t^9 + D_2.t^8 + D_3.t^7 + D_4.t^6 + D_5.t^5 + D_6.t^4 + D_7.t^3 + D_8.t^2 + D_9.t + f_y$$

Where: D_1, D_2, D_3, D_4, D_5, D_6, D_7, D_8 и D_9 are coefficients.

The coefficients should be selected depending on the normatively defined safety factor for the structural steel element.

Chapter V. Implementation of the research findings

The research conducted was applied to three applications (in construction sites).

Application 1. Application of the study to corroded reinforcing steel

A common problem in construction practice is that before the reinforcement is concreted, corrosion is already present (fig. 5.1).

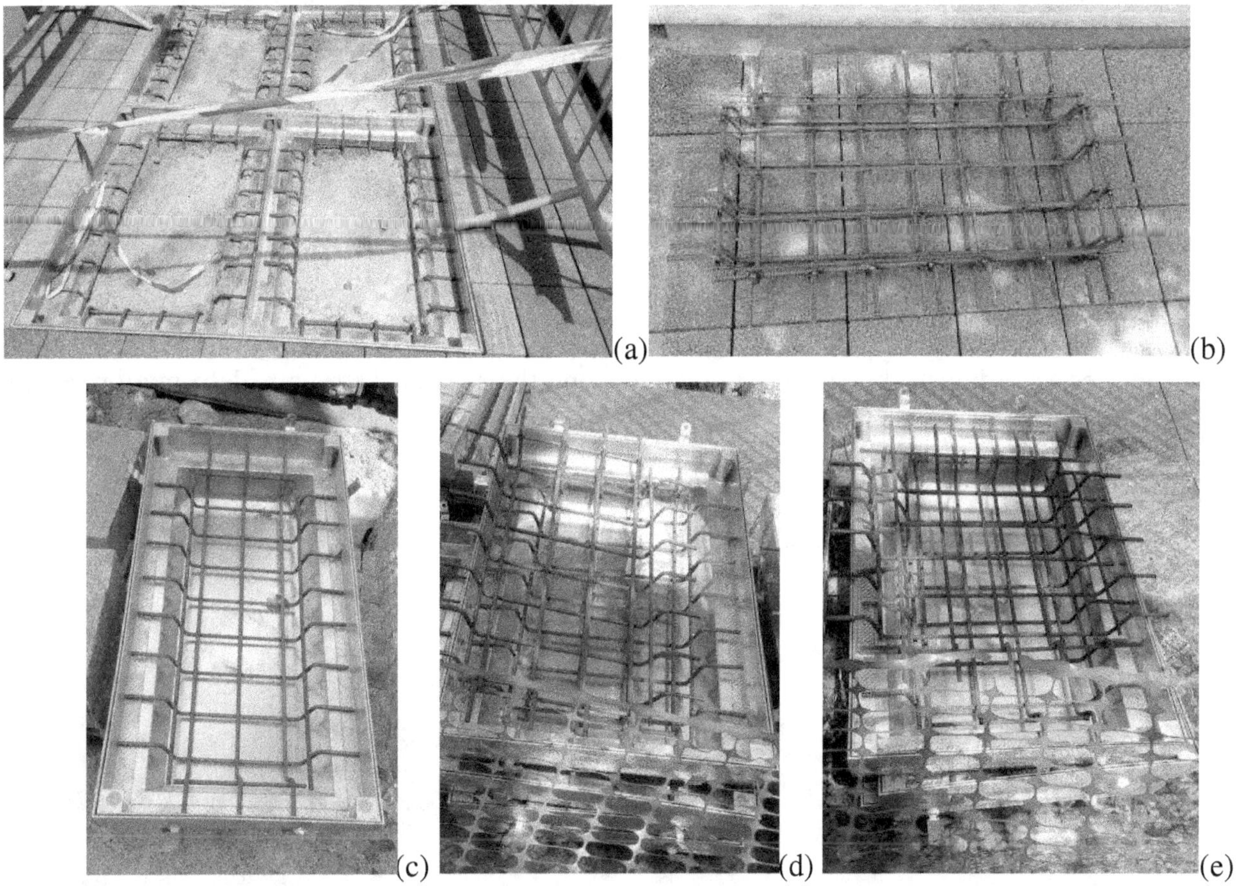

Figure 5.1 Corroded reinforcement at the time of placement; (a) During concreting; (b) Reinforcing mesh before placement; (c) Reinforcement placed in formwork; (d) Placed reinforcement; (e) Reinforcement placed in formwork

Application 2. Application of the study using a corroded steel truss for re-use

There are enough corroded ready-made steel trusses that for various reasons have not been installed, but have been left in storage (fig. 5.2). As a result of environmental influences, whether they have started to corrode and their reuse should be checked using the relationships derived in the book.

Figure 5.2 Corroded steel truss due to its chanting, which will be re-use

Application 3. Application of the study in assessing the possibilities for continued of a corroded steel structure

There are many steel structures available, which for various reasons have not been put into operation and/or have been left to the effects of the environment (fig. 5.3). Calculations are made for the steel structures according to the research in the book and their residual strength, deformation (essential when providing a structure with earthquake resistance) are determined and their residual load-bearing capacity over time is determined. Financial resources of up to 35% are saved for building the supporting structure anew, and from there the cost is reduced.

Figure 5.3 Corroded steel structure for which the residual strength, deformation and residual load-bearing capacity have been determined - the structure will be used without dismantling.

References

[1] ALMUSALLAM, Abdullah A. Effect of degree of corrosion on the properties of reinforcing steel bars. *Construction and Building Materials*, 2001, 15.8: 361-368.

[2] Mohammad M. Kashani, Adam J. Crewe, Nicholas A. Alexander. Stress-Strain Response of Corroded Reinforcing Bars under Monotonic and Cyclic Loading. 15 WCEE, Lisboa, 2012.

[3] ZHU, Wenjun; FRANÇOIS, Raoul. Effect of corrosion pattern on the ductility of tensile reinforcement extracted from a 26-year-old corroded beam. *Adv Concr Constr*, 2013, 1.2: 121-37.

[4] ZHU, Wenjun. Effect of corrosion on the mechanical properties of the corroded reinforcement and the residual structural performance of the corroded beams. 2014. PhD Thesis. INSA de Toulouse.

[5] XU, Yidong. The Corrosion Characteristics and Tensile Behavior of Reinforcement under Coupled Carbonation and Static Loading. *Materials*, 2015, 8.12: 8561-8577..

[6] DEL ANGEL, Francisco Casanova. Fractal Effect of Corrosion on Mechanical Behavior of Unprotected Structural Steel. In: *Developments in Corrosion Protection*. InTech, 2014.

[7] APOSTOLOPOULOS, C. A.; PAPADAKIS, V. G. Consequences of steel corrosion on the ductility properties of reinforcement bar. *Construction and Building Materials*, 2008, 22.12: 2316-2324.

[8] RANJI, Ahmad Rahbar; ZAKERI, Amir Hamed. Mechanical properties and corrosion resistance of normal strength and high strength steels in chloride solution. *Journal of Naval Architecture and Marine Engineering*, 2011, 7.2: 94-100.

[9] PAIK, Jeom Kee; LEE, Jae Myung; KO, Man Ju. Ultimate shear strength of plate elements with pit corrosion wastage. *Thin-Walled Structures*, 2004, 42.8: 1161-1176.

[10] FURUKAWA, Seiji; CHUN, Pang-jo. Experimental and numerical analysis of corroded steel plates subjected to compression buckling load.

[11] OSZVALD, Katalin. Finite element analysis of corroded steel angles under compression. In: *Conference of Junior Researchers in Civil Engineering, Budapest, Hungary*. 2012.

[12] AHMMAD, Md Mobesher; SUMI, Y. Strength and deformability of corroded steel plates under quasi-static tensile load. *Journal of marine science and technology*, 2010, 15.1: 1-15.

[13] HILTI Corrosion handbook, 2015, https://www.hilti.pt/medias/sys_master/h34/h30/9157274632222/Hilti_Corrosion-Handbook_EN.pdf, видян на 20.06.2017 г.

[14] BAJAJ, Srikanth. Effect of corrosion on physical and mechanical properties of reinforced concrete. 2012. PhD Thesis. University of Akron.

[15] APOSTOLOPOULOS, Ch Alk; PAPADOPOULOS, M. P.; PANTELAKIS, Sp G. Tensile behavior of corroded reinforcing steel bars BSt 500 s. *Construction and building Materials*, 2006, 20.9: 782-789.

[16] APOSTOLOPOULOS, Charis A.; DEMIS, Sotiris; PAPADAKIS, Vagelis G. Chloride-induced corrosion of steel reinforcement–Mechanical performance and pit depth analysis. *Construction and Building Materials*, 2013, 38: 139-146.

[17] CHEN, Gang, et al. Experimental study on the properties of corroded steel fibres. *Construction and Building Materials*, 2015, 79: 165-172.

[18] BEAULIEU, L.-V.; LEGERON, F.; LANGLOIS, S. Compression strength of corroded steel angle members. *Journal of Constructional Steel Research*, 2010, 66.11: 1366-1373.

[19] SHETTY, Akshatha; VENKATARAMANA, Katta; NARAYAN, K. S. Effect of corrosion on flexural bond strength. *Journal of Electrochemical Science and Engineering*, 2014, 4.3: 123-134.

[20] Tatsumasa KAITA, Hiroyuki IKEDA, J.M.R.S. Appuhamy, Mitao OHGA, Katashi FUJII, Experimental study on remaining tensile strength of wide steel specimens with different corrosion conditions, *Internet Journal of Society for Social Management Systems*, Issue 6, Vol.1.

[21] FERNANDEZ, Ignasi; BAIRÁN, Jesús Miguel; MARÍ, Antonio R. Corrosion effects on the mechanical properties of reinforcing steel bars. Fatigue and σ–ε behavior. *Construction and Building Materials*, 2015, 101: 772-783.

[22] SUPRUNCHUK, V. K., et al. A study of the effect of corrosion on mechanical properties of steels St. 2 and St. 3. *Materials Science*, 1967, 3.4: 301-303.

[23] APPUHAMY, J. M. R. S., et al. Estimation of corrosion-induced strength deterioration of steel bridge plates—an analytical method. *Annual Research Journal of SLSAJ*, 2001, 11: 19-25.

[24] SHETTY, Akshatha, et al. Performance enhancement of TMT rebar in accelerated corrosion. *Journal of Civil Engineering Research*, 2012, 2.1: 14-17.

[25] HOU, Yingbo, et al. Experimental investigation on corrosion effect on mechanical properties of buried metal pipes. *International Journal of Corrosion*, 2016, 2016.

[26] PONJAYANTHI, D.; VINODH, K. R. Effect of corrosion on the mechanical properties of steel reinforcement. *International Journal of Current Research and Review*, 2016, 8.12: 14.

[27] LEE, Han-Seung; CHO, Young-Sang. Evaluation of the mechanical properties of steel reinforcement embedded in concrete specimen as a function of the degree of reinforcement corrosion. *International Journal of Fracture*, 2009, 157.1: 81-88.

[28] JURIŠIĆ, Paul; PARUNOV, Joško. Influence of corrosion-related degradation of mechanical properties of shipbuilding steel on collapse strength of plates and stiffened panels. *Towards Green Marine Technology and Transport*, 2015, 427-432.

[29] QIN, Guang-chong, et al. Study on the degradation of mechanical properties of corroded steel plates based on surface topography. *Journal of Constructional Steel Research*, 2016, 125: 205-217.

[30] CHEN, Y. Y., et al. Corrosion resistance and mechanical properties of low-alloy steels under atmospheric conditions. *Corrosion Science*, 2005, 47.4: 1001-1021.

[31] FARGAS, G.; ANGLADA, M.; MATEO, A. Effect of the annealing temperature on the mechanical properties, formability and corrosion resistance of hot-rolled duplex stainless steel. *Journal of materials processing technology*, 2009, 209.4: 1770-1782.

[32] DENG, Bo, et al. Effect of thermal cycles on the corrosion and mechanical properties of UNS S31803 duplex stainless steel. *Corrosion Science*, 2009, 51.12: 2969-2975.

[33] ALBRECHT, P., et al. Remaining fatigue strength of corroded steel beams. In: *International Association for Bridge and Structural Engineering Workshop*. 1990. p. 71-84.

[34] LI, J. F., et al. Influence of retrogression temperature and time on the mechanical properties and exfoliation corrosion behavior of aluminium alloy AA7150. *Materials characterization*, 2009, 60.11: 1334-1341.

[35] FRANÇOIS, Raoul; KHAN, Inamullah; DANG, Vu Hiep. Impact of corrosion on mechanical properties of steel embedded in 27-year-old corroded reinforced concrete beams. *Materials and structures*, 2013, 46.6: 899-910.

[36] OK, Duo; PU, Yongchang; INCECIK, Atilla. Computation of ultimate strength of locally corroded unstiffened plates under uniaxial compression. *Marine Structures*, 2007, 20.1: 100-114.

[37] AMLEH, Lamya; GHOSH, Alaka. Modeling the effect of corrosion on bond strength at the steel–concrete interface with finite-element analysis. *Canadian Journal of Civil Engineering*, 2006, 33.6: 673-682.

[38] LI, Fumin, et al. Numerical Simulation Method for Fracture Effect of Corroded Steel Strand Under Tension. 2014.

[39] SANDAMALI, W. K. N.; WIJESENA, H. P.; APPUHAMY, J. M. R. S. Prediction of Residual Buckling Strength in Corroded Steel Bridge Members. 2015.

[40] GENEL, Kenan; DEMIRKOL, Mehmet; GÜLMEZ, Turgut. Corrosion fatigue behaviour of ion nitrided AISI 4140 steel. *Materials Science and Engineering: A*, 2000, 288.1: 91-100.

[41] CINITHA, A.; UMESHA, P. K.; IYER, Nagesh R. An overview of corrosion and experimental studies on corroded mild steel compression members. *KSCE Journal of Civil Engineering*, 2014, 18.6: 1735-1744.

[42] OKAZAKI, Yoshimitsu, et al. Corrosion resistance and corrosion fatigue strength of new titanium alloys for medical implants without V and Al. *Materials Science and Engineering: A*, 1996, 213.1-2: 138-147.

[43] OLIVEIRA, F., et al. Corrosion–fatigue properties of a 4340 steel coated with Colmonoy 88 alloy, applied by HVOF thermal spray. *Surface and Coatings Technology*, 2001, 140.2: 128-135.

[44] MURTAZA, G.; AKID, R. Empirical corrosion fatigue life prediction models of a high strength steel. *Engineering Fracture Mechanics*, 2000, 67.5: 461-474.

[45] CHITTY, J., et al. Influence of electroless Ni-P deposits on the corrosion-fatigue properties of an AISI 1045 steel. *Thin solid films*, 1997, 308: 430-435.

[46] APOSTOLOPOULOS, Ch Alk. Mechanical behavior of corroded reinforcing steel bars S500s tempcore under low cycle fatigue. *Construction and Building Materials*, 2007, 21.7: 1447-1456.

[47] AZAR, V.; HASHEMI, B.; YAZDI, Mahboobeh Rezaee. The effect of shot peening on fatigue and corrosion behavior of 316L stainless steel in Ringer's solution. *Surface and Coatings Technology*, 2010, 204.21: 3546-3551.

[48] PREVÉY, Paul S.; CAMMETT, John T. The influence of surface enhancement by low plasticity burnishing on the corrosion fatigue performance of AA7075-T6. *International Journal of Fatigue*, 2004, 26.9: 975-982.

[49] SHOPOV, Antonio; BONEV, Borislav. Study by experimental of the zone of fracture on S355JR steel specimens with corrosion. *International Journal of Civil Engineering and Technology*, 2019, 10.2: 751-760.

[50] KASHANI, Mohammad M.; CREWE, Adam J.; ALEXANDER, Nicholas A. Nonlinear cyclic response of corrosion-damaged reinforcing bars with the effect of buckling. *Construction and Building Materials*, 2013, 41: 388-400.

[51] LUTES, L. D., et al. Assessing the compressive strength of corroded tubular members. *Applied ocean research*, 2001, 23.5: 263-268.

[52] ARSENIEV, Dmitry G.; IVANOV, Vladimir M.; KORENEVSKY, Maxim L. Adaptive Stochastic Methods: In Computational Mathematics and Mechanics. Walter de Gruyter GmbH & Co KG, 2018.

[53] LASEBIKAN, B. A.; AKISANYA, A. R.; DEANS, W. F. Autoclave design for high pressure-high temperature corrosion studies. *Journal of Engineering, Design and Technology*, 2015, 13.4: 539-555.

[54] AVELDAÑO, Raquel R.; ORTEGA, Néstor F. Behavior of concrete elements subjected to corrosion in their compressed or tensed reinforcement. *Construction and Building Materials*, 2013, 38: 822-828.

[55] LI, Fumin; QU, Yaxu; WANG, Jianghao. Bond life degradation of steel strand and concrete under combined corrosion and fatigue. *Engineering Failure Analysis*, 2017.

[56] WAHAB, M. A.; SAKANO, M. Corrosion and biaxial fatigue of welded structures. *Journal of materials processing technology*, 2003, 143: 410-415.

[57] KIM, In-Tae, et al. Effect of corrosion on the tension behavior of painted structural steel members. *Journal of Constructional Steel Research*, 2017, 133: 256-268.

[58] ПОПОВ, Александър: *Вероятности методи при оценяване на механични свойства на железовъглеродни стомани*. Институт по механика при БАН, 2015.

[59] EVANS, Lawrence C. *An introduction to stochastic differential equations*. American Mathematical Soc., 2012.

[60] KLEBANER, Fima C. *Introduction to stochastic calculus with applications*. World Scientific Publishing Company, 2012.

[61] NAKAI, Tatsuro; MATSUSHITA, Hisao; YAMAMOTO, Norio. Effect of pitting corrosion on strength of web plates subjected to patch loading. *Thin-walled structures*, 2006, 44.1: 10-19.

[62] SHENG, Jie; XIA, Junwu. Effect of simulated pitting corrosion on the tensile properties of steel. *Construction and Building Materials*, 2017, 131: 90-100.

[63] BALUSAMY, T.; KUMAR, Satendra; NARAYANAN, TSN Sankara. Effect of surface nanocrystallization on the corrosion behaviour of AISI 409 stainless steel. *Corrosion Science*, 2010, 52.11: 3826-3834.

[64] TREFETHEN, Lloyd N. *Approximation theory and approximation practice*. Siam, 2013.

[65] BARILE, C.; CASAVOLA, C.; PAPPALETTERE, C. Corrosion effects on mechanical properties of sintered stainless steels. *Procedia Structural Integrity*, 2017, 5: 195-201.

[66] KARAGAH, Hossein, et al. Experimental investigation of short steel columns with localized corrosion. *Thin-walled structures*, 2015, 87: 191-199.

[67] ЦЕНОВ, Цветан. *Корозия на материалите*. Албатрос. 2006.

[68] AHMAD, Zaki. *Principles of corrosion engineering and corrosion control*. Elsevier, 2006.

[69] XU, Fangyuan, et al. Experimental Study on Corrosion and Mechanical Behavior of Main Cable Wires Considering the Effect of Strain. *Materials*, 2019, 12.5: 753

[70] RAMKUMAR, K. Devendranath, et al. Investigations on the microstructure, tensile strength and high temperature corrosion behaviour of Inconel 625 and Inconel 718 dissimilar joints. *Journal of Manufacturing Processes*, 2017, 25: 306-322.

[71] BABOIAN, Robert. *Corrosion tests and standards: application and interpretation*. ASTM International, 2005.

[72] JONES, Russell H. (ed.). *Environmental effects on engineered materials*. CRC Press, 2001.

[73] SHI, Cheng, et al. Numerical investigation of H-shaped short steel piles with localized severe corrosion. *Engineering structures*, 2014, 73: 114-124.

[74] CERIT, Muhammet. Numerical investigation on torsional stress concentration factor at the semi elliptical corrosion pit. *Corrosion Science*, 2013, 67: 225-232.

[75] КИСЛЯКОВ, Сава. *Съпротивление на материалите*. Техника, 1976.

[76] PODSKREBKO, M. D. *Soprotivlenie materialov. Strength of materials*. Moscow, Vysshaya shkola Publ, 2007.

[77] AHN, Jin-Hee, et al. Residual compressive strength of inclined steel tubular members with local corrosion. *Applied Ocean Research*, 2016, 59: 498-509.

[78] LEYGRAF, Christofer, et al. *Atmospheric corrosion*. John Wiley & Sons, 2016.

[79] MARCUS, Philippe (ed.). *Corrosion mechanisms in theory and practice*. CRC press, 2011.

[80] POWELL, Michael James David. *Approximation theory and methods*. Cambridge university press, 1981.

[81] SHOPOV, Antonio; BONEV, Borislav. Experimental study of zone of yield strength on corroded construction steel specimens for reuse. In: *MATEC Web of Conferences 279*. EDP Sciences, 2019. p. 02009.

[82] APOSTOLOPOULOS, C. A.; PAPADOPOULOS, M. P. Tensile and low cycle fatigue behavior of corroded reinforcing steel bars S400. *Construction and Building Materials*, 2007, 21.4: 855-864.

[83] OSZVALD, Katalin; TOMKA, Pál; DUNAI, László. The remaining load-bearing capacity of corroded steel angle compression members. *Journal of Constructional Steel Research*, 2016, 120: 188-198.

[84] KŘIVÝ, Vít; URBAN, Viktor; KUBZOVÁ, Monika. Thickness of Corrosion Layers on Typical Surfaces of Weathering Steel Bridges. *Procedia Engineering*, 2016, 142: 56-62.

[85] PINKUS, Allan. *N-widths in Approximation Theory*. Springer Science & Business Media, 2012.

[86] BRAESS, Dietrich. *Nonlinear approximation theory*. Springer Science & Business Media, 2012.

[87] JONES, Russell H., et al. (ed.). *Stress-Corrosion Cracking, Materials performance and evaluation*. ASM international, 2017.

[88] SYED, S. Atmospheric corrosion of materials. *Emirates Journal for Engineering Research*, 2006, 11.1: 1-24.

[89] SHOPOV, Antonio. Stochastic way for calculation of strength on construction steel with corrosion. In: *XVIII Anniversary International Scientific Conference by Construction and Architecture "VSU'2018", Sofia, Bulgaria,* 2018, 1.1: 413-418.

[90] CHEN, Hui, et al. Experimental Investigation into Corrosion Effect on Mechanical Properties of High Strength Steel Bars under Dynamic Loadings. *International Journal of Corrosion*, 2018, 2018.

[91] SHOPOV, Antonio; BONEV, Borislav. Change of young's module on steel specimens with corrosion by experiment. *International Journal of Modeling and Optimization*, 2019, 9: (in press).

[92] KOCAŃDA, Dorota, et al. Fatigue Behaviour of S235JR Steel after Surface Frictional-Mechanical Treatment in Corrosive Environment. In: *Key Engineering Materials*. Trans Tech Publications, 2014, 105-112.

[93] KULICKI, J. M., et al. Guidelines for evaluating corrosion effects in existing steel bridges. 1990.

[94] APOSTOLOPOULOS, Ch, et al. Characteristic defects-corrosion damage and mechanical behavior of dual phase rebar. *Materials Physics & Mechanics*, 2017, 30.1.

[95] JIA, Chen, et al. Incipient corrosion behavior and mechanical properties of low-alloy steel in simulated industrial atmosphere. *Construction and Building Materials*, 2018, 187: 1242-1252.

[96] LEYGRAF, Christofer, et al. *Atmospheric corrosion*. John Wiley & Sons, 2016.

[97] SEMICONDUCTOR, O. N. LM317 datasheet. 2016.

[98] RADER, David J. *Deterministic operations research: models and methods in linear optimization*. John Wiley & Sons, 2010.

[99] DAVIS, Mark HA. *Markov models & optimization*. Routledge, 2018.

[100] CASSANDRAS, Christos G.; LYGEROS, John. *Stochastic hybrid systems*. CRC Press, 2006.

[101] HARRIS, Jonathan M. Basic principles of sustainable development. *Dimensions of Sustainable Developmnet*, 2000, 21-41.

[102] GRIFFITH, Alan Arnold. VI. The phenomena of rupture and flow in solids. Philosophical transactions of the royal society of london. Series A, containing papers of a mathematical or physical character, 1921, 221.582-593: 163-198.

[103] HYNDMAN, Rob J.; ATHANASOPOULOS, George. *Forecasting: principles and practice*. OTexts, 2018.

[104] KIM, Jae Kwang; SHAO, Jun. *Statistical methods for handling incomplete data*. Chapman and Hall/CRC, 2013.

[105] АПОСТОЛОВ, А. *Разработване на проекти за устойчиво развитие*. Проджекта, София, 2005.

[106] КАЛИНОВ, Красимир. *Теория на вероятностите и статистика*. Нов Български Университет, 2002.

[107] SHOPOV, A. Computation on Corrosion Influence in Ultimate Strength and Strain Depending on Time. International Journal of Mechanical and Mechatronics Engineering, 19.3, 2019, pp. 112-120

[108] SHOPOV, A. Calculation on Yield Strain Depending on Time of Corrosion Influence. International Journal of Innovative Technology and Exploring Engineering, 8.7, 2019, pp. 2391-2396

[109] SHOPOV, A. and Ganev, Sv. "Forecasting the change of yield stress point on steel with corrosion according to corrosion influence for military shelters, In: 9-th International Scientific Conference on Defensive Technologies "OTEH 2020", Belgrad, Serbia, 2020, pp. 469-474.

[110] SHOPOV, A. Theoretical-Computation Conception for Forecasting on Corrosion Influence into Steel Elements at Sustainable Development. International Journal of Innovative Technology and Exploring Engineering, 8.8, 2019, pp. 2253-2261

www.ingramcontent.com/pod-product-compliance
Lightning Source LLC
LaVergne TN
LVHW082343200926
848481LV00012B/3438
* 9 7 9 8 2 3 1 6 2 6 7 9 3 *